Advances in Industrial Control

T0073516

Other titles published in this Series:

Sandor Markon, Hajime Kita, Hiroshi Kise
and Thomas Bartz-Beielstein

Control of Traffic Systems in Buildings

With 101 Figures

 Springer

Sandor Markon, Ph. D. Dipl. Ing.
FUJITEC Co., Ltd.
Product Development HQ
Big Wing
Hikone
Shiga 522-8588
Japan

Hajime Kita, Dr. Eng.
Academic Center for Computing
  and Media Studies
Kyoto University
Yoshida Nihonmatsu-cho
Sakyo-ku
Kyoto, 606-8501
Japan

Hiroshi Kise, M. Eng., Dr. Eng.
Department of Mechanical
  and System Engineering
Kyoto Institute of Technology
Matsugasaki Sakyo
Kyoto, 606-8585
Japan

Thomas Bartz-Beielstein, Dr. rer. nat.
Chair of Algorithm Engineering
  and Systems Analysis
Department of Computer Science
University of Dortmund
D-44221 Dortmund
Germany

British Library Cataloguing in Publication Data
Control of traffic systems in buildings : applications of
  modern supervisory and optimal control. - (Advances in
  industrial control)
    1.Buildings - Mechanical equipment - Automatic control
    2.Automation - Human factors 3.Traffic engineering - Data
    processing
    I.Markon, Sandor A.
    690.2

Advances in Industrial Control series ISSN 1430-9491
ISBN-13: 978-1-84996-604-7   e-ISBN 978-1-84628-449-6   Printed on acid-free paper

The figures in Chapters 2, 3, 15, and 18 are provided by courtesy of Fujitec Co., Ltd., and reproduced with
permission.

Printed in Germany

9 8 7 6 5 4 3 2 1

Springer Science+Business Media
springer.com

Advances in Industrial Control

Series Editors

Professor Michael J. Grimble, Professor of Industrial Systems and Director
Professor Michael A. Johnson, Professor (Emeritus) of Control Systems
and Deputy Director

Industrial Control Centre
Department of Electronic and Electrical Engineering
University of Strathclyde
Graham Hills Building
50 George Street
Glasgow G1 1QE
United Kingdom

Series Advisory Board

Professor E.F. Camacho
Escuela Superior de Ingenieros
Universidad de Sevilla
Camino de los Descobrimientos s/n
41092 Sevilla
Spain

Professor S. Engell
Lehrstuhl für Anlagensteuerungstechnik
Fachbereich Chemietechnik
Universität Dortmund
44221 Dortmund
Germany

Professor G. Goodwin
Department of Electrical and Computer Engineering
The University of Newcastle
Callaghan
NSW 2308
Australia

Professor T.J. Harris
Department of Chemical Engineering
Queen's University
Kingston, Ontario
K7L 3N6
Canada

Professor T.H. Lee
Department of Electrical Engineering
National University of Singapore
4 Engineering Drive 3
Singapore 117576

Professor Emeritus O.P. Malik
Department of Electrical and Computer Engineering
University of Calgary
2500, University Drive, NW
Calgary
Alberta
T2N 1N4
Canada

Professor K.-F. Man
Electronic Engineering Department
City University of Hong Kong
Tat Chee Avenue
Kowloon
Hong Kong

Professor G. Olsson
Department of Industrial Electrical Engineering and Automation
Lund Institute of Technology
Box 118
S-221 00 Lund
Sweden

Professor A. Ray
Pennsylvania State University
Department of Mechanical Engineering
0329 Reber Building
University Park
PA 16802
USA

Professor D.E. Seborg
Chemical Engineering
3335 Engineering II
University of California Santa Barbara
Santa Barbara
CA 93106
USA

Doctor K.K. Tan
Department of Electrical Engineering
National University of Singapore
4 Engineering Drive 3
Singapore 117576

Professor Ikuo Yamamoto
Kyushu University Graduate School
Marine Technology Research and Development Program
MARITEC, Headquarters, JAMSTEC
2-15 Natsushima Yokosuka
Kanagawa 237-0061
Japan

To our families

Series Editors' Foreword

The series *Advances in Industrial Control* aims to report and encourage technology transfer in control engineering. The rapid development of control technology has an impact on all areas of the control discipline. New theory, new controllers, actuators, sensors, new industrial processes, computer methods, new applications, new philosophies..., new challenges. Much of this development work resides in industrial reports, feasibility study papers and the reports of advanced collaborative projects. The series offers an opportunity for researchers to present an extended exposition of such new work in all aspects of industrial control for wider and rapid dissemination.

The interplay between general theoretical methods and the industrial application is a very important part of the field of control engineering. Sometimes the control problem can be described quite succinctly and mathematically and the engineer is able to find an existing mathematical method to solve the problem with little modification. In other cases, the problem posed by the control application leads to new theoretical methods and a new engineering field emerges. But sometimes the solution to the control problem posed by an application is a non-trivial extension of an existing theory–application interplay in this *Advances in Industrial Control* monograph by Sandor Markon, Hajime Kita, Hiroshi Kise and Thomas Bartz-Beielstein. The applications area is that of transportation systems within buildings. This covers a surprising range of systems: elevators, escalators, moving walkways, horizontal elevators, all for human passengers. For cargo transportation there are specialised freight elevators, conveyor systems, autonomous guided vehicles and stacker cranes. These systems are often integral components within larger systems with even nominally autonomous systems often being integrated within larger systems like factory, warehouse or, perhaps not so well known, hospital automation systems.

The volume is divided into four parts with the first covering a description of the various in-building passenger and freight transportation systems. Modelling and simulation techniques occupy the second part of the monograph. As might be expected, the keynotes for this part are sequential, discrete and constrained processes with queue and schedule behaviour to be modelled and simulated.

Part III of the monograph moves on to the methods of intelligent and optimal control. This part opens with an interesting historical preview that discusses the ultimate failure of heuristic rule methods as transportation systems became more technologically complex. This is followed by a review- and tutorial-type chapter on modern optimal control approaches like neural networks, genetic algorithms and combinational optimisation. The application of these techniques to specific in-building transportation systems occupies the remainder of the monograph and forms Part IV of the volume. These chapters report on such systems as elevator control, elevator-group control, multi-car elevator systems, autonomous-guided-vehicle routing control and warehouse scheduling.

Apart from the contributions to the individual in-building transportation control problems found in the various chapters, this monograph draws together the various in-building transportation systems and presents them as a coherent class of technological problems in control engineering. This is a singular achievement and the monograph is a very welcome entry to the *Advances in Industrial Control* series.

M.J. Grimble and M.A. Johnson
Glasgow, Scotland, U.K.

Preface

This book is the result of a loose but long cooperation among the authors. Gradually we came to realize that the problems that we had been trying to make sense of from our different perspectives, are in fact very closely related. Seen in this light, our research results now in turn appear to hold some message for a rather wider audience.

To illustrate what we mean, let us take a step back, and look at our urban life from a birds' eye point of view. We will try to sketch a possible train of thought that might lead to a fresh view of the role of in-building transportation.

Civilization is the history of cities, where people gather, forming creative communities. For any human achievements to arise, merchants, artisans, scientists, artists had to be near to others, so as to share and exchange artifacts and ideas. This was made possible by the emergence of large-scale groups, living in urban settings.

Cities are compact, and this fact could be said to be their *raison d'etre*, as it reduces the need to move people and goods. But there is a limit, and transportation cannot be eliminated altogether. How close we can get to each other is ultimately determined by our physical and psychological properties, and the dimensions of our buildings reflect these constraints.

The height of floors, length of corridors, and so on, are fitted to human scales, and there is not much we can do to reduce these distances further. It is safe to say that we can consider these dimensions as given. Similar arguments based on human factors, like the tolerance of acceleration, or our reaction times, tend to limit the speed of transportation equipment in buildings. Thus at a point, in moving people and goods inside buildings, further improvements must come from smarter use of existing resources, and not from brute force.

This book is about the struggle to conquer this "last mile", or rather last hundred meters, of the distances separating people, and of the rapid advances that have happened in the last two decades in control engineering for building transportation systems.

Many techniques that were developed in this field are rather generic, and one of our goals is to make them accessible to a wider audience. This is done in the hope that by adapting these methods, better controls might solve such problems that otherwise one would be tempted to attack by throwing more resources at them.

The general structure of the book is as follows:

- In Part I, we review very briefly the "plants", that is, the systems that we need to control. This part is not making any pretentions at being an introduction to the system design of elevators or other transportation systems; for that, we recommend some excellent existing books. We are just trying to whet the appetite of the academic community to this fascinating but rather neglected field of study.
- In Part II, we give an overview of a few modeling and simulation techniques that form a necessary background for experimenting with modern control methods. Here again we are not trying to make a complete review; instead, we just give a glimpse of a few, rather subjectively selected, topics; mostly those that we have actually found important in our work.
- With Part III we arrive at the central topic of this work: introducing the control methods that are revolutionizing our subject field. We have tried to collect not only the established "state-of-the-art" techniques, but also some experimental and controversial ones. We hope that the reader will find the exposition usable and maybe thought provoking.
- Part IV is again a necessarily incomplete section; we have picked some specific topics from our research work that could help the reader see how the methods of Part III are used in practice. The examples range from almost completely theoretically oriented ones, through proposed solutions to practical problems, to actual implemented methods that power commercial systems.

Our hope is that this book could be a bridge between researchers, who might want to try their hand at this interesting field; and practiting engineers, who might find just the right solution to the difficult control problems that they encounter in their work.

Finally, the authors wish to thank the many people and organizations, whose support and help have made possible our research.

H. Kita and S. Markon want to thank first our teacher, Dr. Yoshikazu Nishikawa, President of Osaka Institute of Technology, who has kindled our interest in this field.

S. Markon has also enjoyed the full support of his company, Fujitec Co., Ltd., and in particular, of its founder and former President, the late Mr. Shotaro Uchiyama, the current President, Mr. Takakazu Uchiyama, its Chairman, Dr. Kenji Otani, and the director of engineering, Mr. Akira Sumimoto. Of the many others to whom we are indebted, a small sampling is: Prof. Shinzo Kitamura and his group of Kobe University; Prof. Mitsuhiko Araki and his group of Kyoto University; Prof. Hans-Paul Schwefel and his group of Dortmund University; Prof. Kotaro Hirasawa and his associates of Waseda University; and Mr. Kenji Sasaki of Fujitec Co., Ltd.

H. Kita is further indebted to Prof. Shigenobu Kobayashi and Prof. Isao Ono of Tokyo Institute of Technology, Mr. Yoshiaki Komoriya, Mr. Satoshi Takahashi, Dr. Yasuhito Sano, Mr. Hiromichi Suzuki for collaborative study in evolutionary computation. He also thanks Prof. Hans-Paul Schwefel of Dortmund University and his group for acceptance as a visiting researcher and valuable discussion.

H. Kise expresses his gratitude to Dr. and Prof. Emeritus Minoru Uno of Kyoto Institute of Technology, Dr. and Prof. Emeritus Hisashi Mine of Kyoto University and Dr. and Prof. Toshihide Ibaraki of Kansei Gakuin University. They have always encouraged him since he started his study 35 years ago. He also thanks his wife Yoshiko for her warm support.

T. Bartz-Beielstein expresses his thanks to Prof. Hans-Paul Schwefel and Prof. Günter Rudolph of Dortmund University who provided a productive working atmosphere. Prof. Thomas Bäck of Leiden University established his contact with S. Markon. The cooperation with Prof. Dirk Arnold of Dalhousie University and Mike Preuß and Christian Lasarczyk of Dortmund University produced valuable scientific results.

Kyoto,
January 2006   *S. Markon, H. Kita, H. Kise, and T. Bartz-Beielstein*

Contents

Part IV Topics in Modern Control for Transportation Systems

Part I

Transportation Systems

1

Introduction

Transportation systems share an aura of familiarity that probably makes it all the more difficult to deal with them objectively. We tend to have "gut feelings" about railway systems, how the level crossing *should* work and how the timetable *could be* improved. We all "know" that the traffic signals in front of the supermarket are just simply *incompetently* done. I can imagine very vividly the exasperated traffic engineers, trying to explain some basic facts about their discipline to the public.

These observations seem to be doubly so for transportation systems in buildings. In particular, elevators are deliberately designed to be unobtrusive, familiar looking, and simple. Many people are genuinely surprised hearing for the first time that there *is* such a thing as "elevator group control". I am sure most elevator control engineers have had their share of otherwise smart and educated people, telling them as a revelation the hoary old anecdote of the miracle cure for long waits: "Just put a mirror in the elevator hall!".

We ask the reader to put aside any preconceptions, forget what he thinks he knows about buildings, even if a good part of his life is spent there. The systems and control methods that we are going to survey here are very different in reality from the superficial image that most people form in everyday life.

When we are talking about transportation systems in buildings, it is useful to consider their primary differences from other transportation systems.

- **Fast, real-time decisions**
  Here we deal only with relatively short distances, so the time scale is rather dense. Control decisions are made in real time, on the second or subsecond order; we should contrast this with other public transportation systems, that might use precomputed timetables, or make "real-time" decisions on the order of hours, like opening or closing road lanes.
- **Individualized bulk service**
  Although these systems are public, just like roads or railways, it is not sufficient to consider only the bulk traffic. Individual persons or packages

have to be catered for, in a manner similar to a taxi service; but at the same time these systems have to cope often with enormous traffic volumes.

- **Inherent, essential randomness**
  When we are dealing with individual people, their behavior is predictable only in a weak sense. Thus as a corollary to the previous property, the traffic demand inevitably shows a strong stochastic component, that has to be taken into account for effective control.

We should mention that we are witnessing a gradual convergence of "wide-area" and "local" transportation engineering. For instance, one goal of the important emerging field of *intelligent transportation systems* (ITS) is the introduction of fast real-time control into road traffic systems. In an ITS system, intelligence can be added on several levels.

- **Local level:** Individual vehicles are equipped with sensing, guidance, and information systems, to improve their interaction with other participants (vehicles, pedestrians) of the traffic.
- **Mesoscopic level:** Vehicle-vehicle interactions can give rise to group behavior; one typical example is the *ad hoc* formation of convoys, by linking the speed control systems of cars on the highway.
- **Global level:** Control of the routing and flow rates of the road traffic can be used to achieve higher throughputs, or to dissolve traffic jams, *etc.*

Some elements of such local and global control mechanisms show similarities to the building transportation systems, that are our subject. However, the above differences are still valid for current typical traffic systems.

Returning to the topic of in-building transportation, in an urban setting, people and goods enter and leave buildings around the clock. That implies internal circulation, and thus a transportation system.

We are going to review in very general terms the most common transportation systems used in our city buildings, both from the point of view of how they are built, and how they are used. We need to restrict the discussion to the extent needed for later sections, and the interested reader is referred to the many excellent books dealing in depth with particular systems.

It should be obvious from the above, that the control of building transportation systems is demanding and nontrivial. Fortunately, recent developments in computation, applied mathematics, and information science have made it possible to fundamentally revise the approach to this subject.

The field of optimization has been revolutionized in recent decades. Bioinspired algorithms, that make use of parallelism and randomness belong to the every-day-tools of modern optimization practitioners. These algorithms are applied to tackle problems that cannot be solved with classical algorithms due to:

Noise: Measurement errors prevent the determination of exact function values.

Randomness: Optimization via simulation is based in many cases on stochastic simulation models. Consider *e.g.* the stochastic arrivals of passengers in buildings for elevator group control.

No derivatives: Derivatives, which are required for many classical optimization algorithms, are unknown or cannot be determined.

Multimodality: Many real-world optimization problems have more than one optimum.

Multicriteria: Time against costs is only one example of conflicting goals that lead to multicriteria optimization problems.

Dynamically changing systems: The objective function for many real-world problems changes over time.

Modern optimization methods, capable of dealing with transportation problems having the above characteristics, are introduced in Part III, while some actual applications are reviewed in Part IV.

2

Passenger Transportation Systems

Helping persons move around the building is the primary function of the transportation infrastructure. There are several difficult requirements that have to be satisfied by any passenger transportation system, but the first one is *safety*. Potential dangers to passengers in buildings involve falling, crushing, getting trapped, and many other possibilities. People must be protected not only from equipment malfunctions and other accidents with external causes, but also from the unintended consequences of their own actions, whether due to carelessness, or deliberate misuse.

The invention of the *elevator safety device* by Elisha G. Otis, which he dramatically demonstrated in 1854 by cutting the suspending rope over his head [1], is seen as one of the crucial technical advances that made possible the birth of the *high-rise building*, and thus the modern metropolis (see Fig. 2.1).

However, this was only the beginning; today, modern elevators and other transportation systems have far surpassed that stage, and now employ a wide variety of other devices and design features for exceptional safety.

The modern transportation equipment has evolved mostly on the principle of making in-building travels ever safer, and pursuing efficiency and economy only when safety is assured. One piece of evidence for this is the gradual disappearance of such equipment that was once popular and widely used, but now their safety is no longer considered sufficient. An example is the *paternoster* elevator.

The paternoster works on the principle of the *ferris wheel*: its open cabins move continuously, up on the left side and down on the right. Periodically, one of the cabins will appear behind an opening for a short time, allowing a passenger to step in, assuming a certain nimbleness. The passenger then rides the cabin until arriving at the destination floor, where he will step out briskly. There is some amount of mechanical provision to prevent crushing, by an arrangement of moving flaps on both sides of the gap between cabins and floors; but generally it is a better idea not to get caught between the cabin and landing in the first place.

Fig. 2.1. An elevator group at the "Kolon Tower", Kwa Chon, Korea

At first glance, the idea of continuously moving cabins is appealing by its simplicity and the potential for high transportation capacity with no waiting time; however, experience has shown that this system can be dangerous, and new installations have been prohibited in most countries [1].

In the following section we will describe the most prominent member of the passenger transportation family, the elevator, in some detail. Other equipment will be reviewed only cursorily, to give an idea of the options that transportation engineers have.

For more detailed information about all aspects of vertical transportation, we recommend the works of Strakosch [2, 3] and Barney [4, 5].

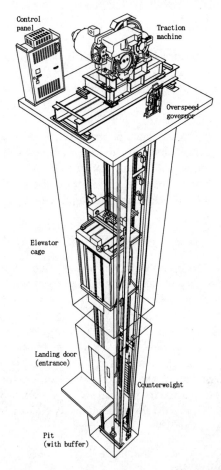

Fig. 2.2. A modern passenger elevator with structural parts shown in the machine room, hoistway, and landings

2.1 Elevators

Elevators are by far the most important transportation systems for handling both passenger and freight traffic in buildings. They provide safe, fast, and economical movement for people and goods, and they are able to cater to all kinds of traffic patterns.

The overall structure of an elevator is shown in Fig. 2.2. Starting from the top, we can see the main parts:

- the *traction machine*, driving the *traction sheave* with an electric motor,
- on the left, the *control panel* that houses the power and control electronics,
- the *elevator cab*, guided between *guiderails*,
- a *landing* with the *landing doors* closed,
- opposite to that, the *counterweight* that balances the weight of the cab.

2.1.1 Construction and Operation

The modern passenger elevator is predominantly of the so-called *traction type*, that uses steel wire ropes wound around rotating sheaves to lift the elevator cabin. Traction is achieved by the friction between the wire ropes and the grooves of the sheave.

In normal usage, the elevator operates automatically, according to the calls registered by users. Its sequence program controls the speed-control electronics, that in turn drives the main motor, to accurately execute the pre-programmed speed profiles, and to position the elevator at the landings with millimeter-order precision. The sequence program also controls the doors, the indicator lights, chimes, and other interface equipment.

2.1.2 Safety

Although wire ropes are designed with a very high safety factor and the occurrence of breakage is extremely rare, the elevator is further protected against falling by multiple safety devices, among them purely mechanical *safety catches* that grab the *guiderails*, when tripped by *overspeed sensors*.

It is generally acknowledged that the traction elevator is the safest and most reliable transportation device, especially when we consider the traffic volumes handled every day. Elevators are used by people of all ages, who in most cases do not even realize that "transportation" is taking place. To the typical user, the elevator appears just as a virtual doorway that happens to open to other floors. This trust by the public is earned by the exceptional safety record of the elevators.

2.1.3 Modern Technology

Technical developments of elevators have continued during the past century, resulting in speeds up to 1000 m min^{-1} (16.6 m s^{-1}) and capacities up to 120 persons. Modern elevators use sophisticated computerized control systems, high-performance materials, and a constant stream of innovation is driven by a fierce competition among the leading manufacturers.

The interested reader is referred to the trade journals, the most prominent among them *Elevator World magazine*, for fascinating accounts of not only the technical achievements, but also of the many social and cultural aspects connected with elevators. Modern aspects of vertical transportation equipment in the context of intelligent building systems is treated *e.g* by So [6].

2.1.4 Control

Elevators are controlled by *calls*: signals from push-buttons operated by the users. We talk of *hall calls*, the signals calling an elevator to a waiting person, and *car calls*, the signal from a person riding the elevator, telling it where to go.

Recently a third kind of call, the *destination call* is getting more and more popular. With destination calls, the waiting person signals the elevator not only their presence and intended direction, but also their destination. This allows more efficient control by grouping passengers by their destinations, thus reducing the stops made by the elevators.

Elevators acknowledge calls by lighting up the *tell-tale lamp* which is usually built into the hall call button, and eventually serve the call. This is however just a rough approximation; details of the interaction between users and elevators depend on the brand, the geographic area, and the profile of the building. Also, with the proliferation of graphical display devices, elevator systems are starting to provide more and more detailed information to users.

For the purposes of this book, the most important feature of elevators is their ability to operate in cooperating groups. By dividing the traffic dynamically among several elevators, it is possible to achieve an increase in total traffic-handling capacity that can significantly surpass the sum of the traffic capacity of isolated individual elevators.

In the rest of this book, we will be concerned with control methods that make possible such "something-for-nothing" effects.

2.2 Other Passenger Transportation Equipment

2.2.1 Escalators

Besides elevators, escalators are the mainstay of vertical traffic-handling equipment.

The escalator, as a mass transportation device, was developed and commercialized at the turn of the 20th century. Escalators provide a continuous, one-way connection between two levels, usually between consecutive floors. They operate on the principle of driving a connected chain of individual *steps*, running on rollers along two inclined loops of guiderails. They are equipped with a pair of handrails, driven synchronously with the steps, for the safety and convenience of the passengers. The structure of a conventional escalator is shown in Fig. 2.4.

Escalators are usually designed for one or two passengers/step (step width of 600–1000 mm), and operate at a speed between 30–45 m min^{-1}. Typical performance figures for escalators are given in [2]; the general range of carrying capacity achieved in practice is reported as about 2000–5400 persons/h.

There are some situations where escalators show clear advantages over elevators, as they can cope with very high, continuous traffic volumes, and there is no perceived waiting time when using them. These properties make them ideal for moving large crowds up or down for one or a few floor levels,

Fig. 2.3. Escalators at the "Miramar Entertainment Park", Taipei, Taiwan

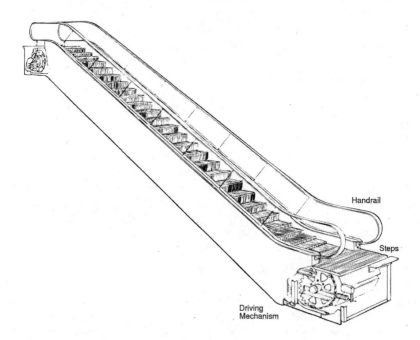

Fig. 2.4. Structure of the escalator. The driving mechanism, shown in breakout view, is embedded in the floor structure

which is the typical requirement in subway stations, airports, and other transit facilities.

They are also often used in department stores and other shopping centers, where they have the additional advantage of allowing passengers a continuous view of the surroundings (see Fig. 2.3). This feature is attractive for the shops, as it can increase the exposure of the sales areas.

Escalators do have, however, some shortcomings that limit their use. They are too slow for long distances (over about 10 to 15 floors), and they cannot be used by everyone, such as persons with physical disabilities, people with baby carriages and other bulky loads, or in any situation where there is a danger of the rider stumbling.

In the study of transportation systems in buildings, escalators can be treated as a first approximation as passive pathways that people traverse with a constant speed.

2.2.2 Moving Walkways

A variant of the escalator is the moving walkway, or horizontal escalator (see Fig. 2.5), which is often constructed similarly to the inclined escalator: as a continuous chain of steps, which are in this case not stair-shaped but flat.

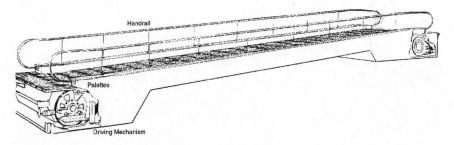

Fig. 2.5. Moving walkway. The driving mechanism, shown in a breakout view, is embedded in the floor structure

There are many variants; for instance, the load-carrying surface can be a rubber belt, running over horizontal rollers. Moving walkways can be inclined by a few degrees, which allows them to connect different concourse levels at airports, allowing people to ride them with baggage carts.

From the point of view of transportation, moving walkways can be treated in the same category as escalators.

2.2.3 Horizontal Elevators

Although in most buildings vertical movements dominate the transportation task, airports present a special case. In airports, the large horizontal distances between gates warranted the development of the horizontal elevator. These *people movers* appear to the public as small-sized short-haul trains, but in fact their technology is mostly borrowed from elevators. They move automatically from station to station in an enclosed driveway, with automatic doors, and their driving method is often the wire rope, just like the usual elevators.

In the future, we can expect horizontal elevators appearing in the urban scene, as ultra large-scale building complexes begin to require a horizontal transportation capability on medium distances.

3

Cargo Transportation Systems

Every building requires the movement of freight, both in and out, and inside the building. For instance, office buildings receive a large amount of mail, food, office supplies, repair parts *etc.* every day; while generating outgoing mail and garbage.

In practice, the horizontal movement is often negligible, and the vertical part dominates. The transportation systems are usually designed to handle this part.

3.1 Freight Elevators

In many buildings, a large part of the cargo traffic is handled by elevators; these are sometimes shared with passengers, but often freight elevators are used when the cargo would not fit physically or logistically. For instance, specialized elevators are used in parking houses to move cars.

Another case is the "dumbwaiter" , often used in restaurants, hotels, and other cases where there is a frequent need to move compact cargo between levels, like the kitchen and dining floor (see Fig. 3.1(b)). These small elevators are designed with such dimensions to make it impossible for people to ride them, for safety reasons.

3.2 Conveyors

Besides elevators, many other kinds of transportation systems have also been developed; for instance, Shinko Electric Co., Ltd. of Japan offers a network-type linear motor system "Super LimLiner", for use in hospitals, that can move packets both horizontally and vertically. Although such systems present a clear potential advantage, at this time we have insufficient experience with them to allow us to treat them from the point of view of transportation control.

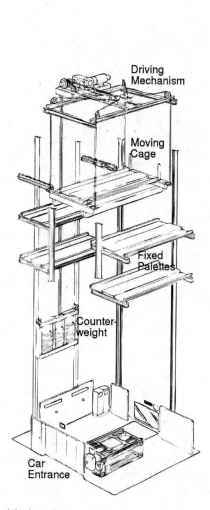

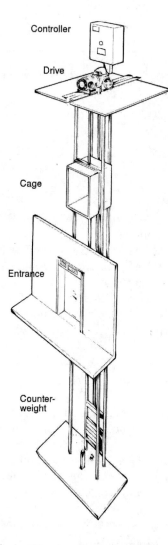

(a) An elevator for automated parking houses. Cars are carried on palettes that slide into the parking slots

(b) A "dumbwaiter" (small-capacity freight elevator). Notice the structure essentially identical to the passenger elevator

Fig. 3.1. Special-purpose elevators

3.3 Automated Guided Vehicles

Automated guided vehicles (AGV) are autonomous electric cars designed for indoor operation. Their most important distinguishing characteristic is that they don't need special guideways unlike conveyors; instead, they can share corridors and elevators with people, with some minimally obtrusive help for guidance. This might involve infrared or radio beacons on the ceilings and painted track markings or magnetic tapes pasted on the floor.

AGVs have been introduced in almost all facilities that need to carry various items between different locations in buildings without guide rails. The most typical usage of AGVs is on the factory floor where AGVs carry items such as parts and tools between machines and warehouses. Modern factory automation (FA) can not be realized without AGVs. However, the transportation problems in FA are intricately connected with efficient scheduling which will be discussed later.

As components of building transportation systems, AGVs are still in the stage of finding niches of usage. One area of clear growth seems to be the case of hospitals. For instance, AGVs form an essential component of the internal transportation system at a hospital in Singapore (see Figs. 3.2(a) and 3.2(b),), where they roam the corridors and ride the elevators, collecting and distributing linen and other items. Similar systems are described by Siikonen in [7].

From our point of view, AGVs are best treated as autonomous agents; they are clearly different from fixed components like elevators or escalators. We shall describe one approach to integrate AGVs into a building transportation system in Chapter 17.

(a) AGVs working at a hospital (one loaded, one free); notice the markings on the floor

(b) AGV riding the elevator (the elevator is called automatically by the AGV control center, and boarded autonomously by the AGV)

Fig. 3.2. AGVs at a hospital

3.4 Stacker Cranes

Stacker cranes are designed to move simultaneously in the horizontal and vertical directions by means of two independent motors (see Fig. 3.3). The most typical usage of stacker cranes is in warehouses where a stacker crane, holding items to be stored, moves in an aisle between two adjacent vertical shelves. It also retrieves items saved in the shelves. So it is also referred to as a storage and retrieval machine (S/R machine). Many stacker cranes are automated by means of an automatic positioning system with computers, and comprise key elements in automated warehouses or automated storage and retrieval systems (AS/RSs) that are components of distribution centers (referred to as depots) for supply chains.

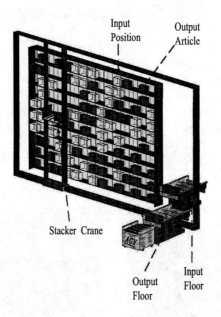

Fig. 3.3. An automated warehouse with stacker cranes

4

External Connections and Related Systems

4.1 External Connections

The building transportation system is mostly self-contained, but there is necessarily some connection with roads, public transportation, or neighboring buildings. This can be in a trivial form of pedestrian passageways or freight loading decks *etc.* However, we see more and more cases where city planning involves an integrated network of moving walks, escalators, and elevators, complementing the traditional passive connections.

4.1.1 Pedestrian Connections

Large volumes of passengers entering and leaving buildings are usually handled by providing sufficient open spaces, with a minimum of features that would disrupt the flow of people. From the point of view of building transportation equipment, these parts of the surroundings serve as buffers that can absorb fluctuations in the flow. This is somewhat different if there are active elements, like escalators or moving walkways; their transport capability will need to be taken into account as a limiting factor on the traffic flow.

4.1.2 Freight Connections

Large buildings have their own road connections for regular truck traffic. From the point of view of transportation, the loading/unloading dock and its associated equipment can be viewed as a source/sink of the freight traffic, with some given traffic volumes and buffer capacity.

4.2 Related Systems

We just briefly mention a few specialized but important areas, where the methods of this book might find direct application.

4.2.1 Factory Automation

Factory automation makes extensive use of specialized transport equipment, and in particular, AGVs. In this context, various AGV configurations and control methods had been studied, and these results will be useful when AGVs are more widely used in other buildings.

There is one AGV topology that has a strong connection with our study of building transportation systems. Kise *et al.* had studied extensively the AGV system operating on a unidirectional ring path. This configuration is a horizontal counterpart of the elevator group, which can also be viewed as running on a circular path (up then down then up...). Thus the methodology and analytic results obtained for this kind of AGV systems should have applicability for elevators and other transport systems (*e.g.* demand buses).

4.2.2 Warehouse Automation

Automated warehouses might use either a combination of elevators, stacker cranes and AGVs, or integrated conveyor systems. In both cases, the control problem is similar to the usual building traffic control, with an additional degree of freedom of choosing storage space for incoming items. This freedom might be used in optimizing some desired system parameters, like *e.g.* maximizing the throughput or minimizing the expectation of the pickup time.

However, the warehouse transportation problem cannot be completely separated from the more fundamental inventory control; thus a combined approach should be advantageous.

4.2.3 Hospital Automation

Hospitals are a rather special kind of building, from the point of view of transportation. There is a very clear priority structure on transportation demands, with an absolute need of instant availability for emergency transfers; a fluctuating traffic of hospital staff, patients, visitors, support personnel; and a scheduled background traffic of material movements.

Hospitals are at the forefront of introducing AGVs into the urban scene, driven partly by high personnel costs. AGVs are used for transporting clinical records, medicine, food, linen, *etc.* A special example of the automated warehouse is a Japanese university hospital that utilizes an automated container-storage management system linked with a reservation system for surgical operations. The container is automatically carried out on the operation day, according to a doctor demand. Currently AGVs and elevators are controlled separately: from the elevators' point of view, AGVs are just passengers; conversely, AGVs see elevators as a (random) delay. This will definitely change in the near future, with the introduction of integrated control.

A study of the authors concerning the cooperative control of AGVs and elevators is introduced in Chapter 17. Another example of a report dealing with related questions is found in [7].

Part II

Modeling and Simulation

5

General Modeling Concepts

In this chapter we will review briefly some properties of the building transportation systems that are relevant to the control task. Our chief interest is to lay the foundation for treating the transportation systems in a quantitative way. For this, we shall introduce the most representative systems from the point of view of giving numerical values to their performance parameters.

5.1 Components and Topology

The main building elements of the transportation system are the vehicles that carry passengers or goods, and the tracks where these vehicles travel.

These components are supported by a signaling, command, and communication system.

5.1.1 Vehicles

Elevator cages are the most representative vehicles in buildings. From the control point of view, their important characteristics are the following:

- Carrying capacity (in persons)
- Traveling speed (floor-to-floor or *brake-to-brake* time)
- Transfer time (including door open/close time, depending on door width)

Some typical values of the most common parameters are shown in Table 5.1.

In a sense, we can speak of the escalator steps also as *vehicles*. They are designed to carry either two persons (in the case of a step width of 1000 mm) or one person (for 600 or 800 mm wide escalators).

Elevator Dynamics

From the point of view of group control, the most important dynamical characteristics of an elevator are

Table 5.1. Common ranges of elevator operating parameters

Item	Lower values	Higher values	Unit
Rated capacity	6	24	persons
Contract speed	0.5	6.0	m s^{-1}
Acceleration/deceleration	0.7	0.9	m s^{-2}
Door opening/closing times	2.0	3.0	s
Passenger transfer times	0.8	1.5	s

1. operating times:
    a) running times $T_{\mathrm{run}}(f_{\mathrm{start}}, f_{\mathrm{stop}})$ between any two floors $f_{\mathrm{start}}, f_{\mathrm{stop}}$ (including acceleration and deceleration times),
    b) door times T_{door} (door open/close time, door holding times),
2. operating sequence,
3. service floors,
4. rated capacity (persons).

The *running times* are determined by the *physical parameters* of the elevator drive, by the dimensions of the building, and by *physiological constraints* on comfortable levels of acceleration. An idealized speed diagram [8] of a traction elevator, based on these data, is shown in Fig. 5.1. In the first approximation, the top speed v_{max} (called *contract speed*) [1], the highest allowed acceleration (deceleration) a_{max}, and the running distance s, together with some corrections for various loss times, determine the running time T_{run}, or equivalently the so-called *brake-to-brake time*. Usually, a_{max} can be considered the essential limiting factor, which cannot be increased over the physiologically determined level of about $0.1g$. The *door times* T_{door} depend partly on the opening width, the number of moving panels (center or side open, *etc.*), and the quality of the door-operating mechanism. For the protection of passengers, the kinetic energy of the closing door has to be limited. The maximum allowed kinetic energy of the door, together with the minimum required mass (for fireresistance etc.) determines the upper limits of the door speed. Additionally, open holding times are determined by the requirement of giving passengers time to reach the door from inside and outside. These in turn depend on *psychological considerations* of the reaction times of passengers waiting for the elevator, and are influenced by the signaling devices employed by the elevator system.

5.1.2 Guideways

Elevator shafts, as guideways, are reserved exclusively for the elevators, and guarded by electric interlocks against possible accidents due to intrusion. In most cases they are completely enclosed, except in the case of panoramic

---

[1] For short distances the contract speed is not reached, and only a_{max} and s matter.

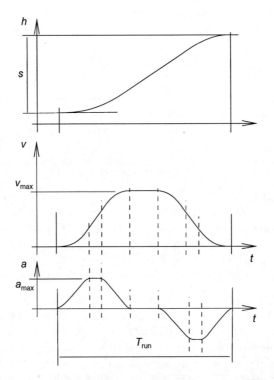

Fig. 5.1. Typical (idealized) speed diagram of a traction elevator car. From top to bottom: running distance s, speed v, and acceleration a, respectively, plotted against time t.

elevators, where hatchways can be partly or fully open to the outside or to an atrium.

Elevators run along sets of steel rails, which are dedicated to one elevator each. This might sound wasteful when compared with roads that are shared with many vehicles.

Indeed, it is likely to change in the future, when new technologies can allow two or more elevator cars to share a common set of rails and shaft space. There are already existing experimental *single-shaft multicar elevators*, and some research results concerning their control are introduced in later chapters.

Elevator shafts are often concentrated into a block, not only for architectural reasons, but also since a group of elevators should be located as close together as possible, to reduce the walking distances for passengers.

From the point of view of transportation engineering, the area taken up by elevator shafts represents a loss of rentable area in the building; thus we are interested in how small it can be made while maintaining the desired level of service. This is the main motivation for the research done in developing single-shaft multi-car elevators.

5.1.3 Signal Systems

Building transportation systems need to provide communication between central or distributed controllers, vehicles, and with individual passengers.

In case of elevators, the communication is typically handled by dedicated serial lines, usually located in the elevator shafts. These connect the elevator controllers, elevator cages, the hall fixtures, and building automation centers.

Elevator Control

The *operating sequence* of the elevator car is usually the *selective collective control* [5] (abbr. S/C). The elevator moves according to passenger requests (calls), registered at the elevator hall *hall call*, and inside the car (*car call*). The concept of *call* is central to elevator control, so it is briefly explained here:

- A *hall call* is the pushbutton command signaling the *request for service*, given to the elevator system by a passenger, who arrives at an elevator landing, and finds no available elevator car. Usually we speak about *up hall calls* and *down hall calls* for up- and down-traveling passengers.
- A *car call* is the pushbutton command signaling the *destination floor*, given in the elevator by the passenger who has boarded the car.

A simplified description of the operating sequence is the following:

1. A free car (that is, a car that currently has no calls) responds to the first call (car call or hall call), sets its operating direction towards that call, and starts running.
2. While there are *forward calls* (calls in the present direction), the car stops sequentially at each call.
3. When there are no more *forward calls*, the car goes to the first *reverse call*; reverses the operating direction, and repeats the above.
4. When there are no more calls, the car stops at the last served call, and waits for the first new call.

An example is shown in Fig. 5.2. Hall calls are indicated by triangles, and car calls by circles. The car is indicated by the grey arrow shape, and it moves according to the S/C sequence. Having just left Floor 1F, it is moving now in the *up* direction; on the way up, it will stop at the hall calls at 2F and 5F (numbered 1 and 4); also, at the car calls at 3F , 4F, 5F (coinciding with the up hall call), and last at 6F. It will then reverse direction to *down*, and goes down, serving the 4F hall call and the 1F car call. Altogether, it will serve the calls in the order of $1 \to 2 \to 3 \to 4 \to 5 \to 6 \to 7$, stopping at 2F $\to$ 3F $\to$

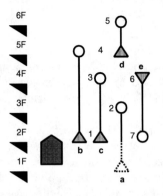

Fig. 5.2. Example of S/C operation

4F → 5F → 6F → 4F → 1F. That means that the passengers are served as a → (c,b) → d → e.

There are other control sequences in use, e.g. the *single automatic* operation, where the car calls originating from each hall call are served before the next hall call is taken, even if that means bypassing hall calls on the way; however, this sequence is very inefficient.

Strictly speaking, the operating sequence of each car should be considered as part of the elevator group control algorithm, and it should be allowed to be optimized. In fact, Tanaka [9, 10] has been pursuing this line of research with promising results. However, users are accustomed to the S/C sequence, which is considered *given* for group control purposes. Also, Closs [11] has found that the S/C control is nearly optimal for the operation of a single elevator car, and other sequences would provide only marginal gains.

To give an example for a control sequence better than the S/C control, it might be possible to improve the efficiency in Fig. 5.2 by serving the calls in a different order; one stop could be saved by the stopping sequence 2F → 3F → 5F → 6F → 4F → 1F, thus letting out passenger c on the downward trip. However, by the current conventional thinking, such situations are not frequent enough to justify abandoning the S/C sequence.

As seen from the above, the lower limits on elevator operating times are determined by physiological, psychological, and physical factors, so improving the elevator system performance is not simply a matter of using faster elevators. Even if the elevator drive mechanism were made more powerful, the physiological and other constraints would prevent substantial improvements in the system performance. On the other hand, better group control methods are not subject to such constraints, and they can be expected to be effective for most elevator systems.

5.1.4 Zones and Banks

From an engineering point of view, it is advantageous for an elevator to serve a sequence of adjacent floors. However, the maximum number of the floors that can be served is limited by the time it takes to travel from bottom to top, stopping at each floor, since the passenger traveling all the way would have to endure this.

This conflict has resulted in the compromise of dividing the building into zones, where each elevator serves only one zone, besides the main lobby.

Elevators serving a zone are arranged as a group, dividing the traffic in the zone among themselves. A typical group of elevators in an office building would be in most cases between 4 to 6 cars, with 8 cars the practical maximum.

Elevator engineers have developed practical rules to estimate the overall performance of banks of elevators. The design calculation is described in detail by Barney [5]; here we just summarize the basic concept for later reference.

As the most simple case, let us consider *up-peak* traffic for a building with one bank of L elevators serving N floors above the lobby floor. As an idealization, we can assume that the elevators arrive at some unknown but equal INT intervals to the main lobby floor, where the same number of P passengers will board each elevator. As the elevator runs upwards, it will make stops to let one or more passengers out, at an expected S floors. Assuming equal probabilities for each destination, S is obtained as

$$S = N \left[1 - \left(\frac{N-1}{N} \right)^P \right] \tag{5.1}$$

With this, we can calculate the *round trip time* RTT as

$$RTT = 2Nt_v + (S-1)t_s + 2Pt_p \tag{5.2}$$

where t_v, t_s, t_p are the running time for one floor, the total stop time (decelerating, door open/close, accelerating) for one stop, and the total transfer time (boarding, leaving) for one passenger. Actual values are obtained partly from the specifications of the equipment (as indicated in Section 5.1.1), and partly from field experience.

With RTT, we obtain the *average interval INT*:

$$INT = RTT/L \tag{5.3}$$

and thus we can estimate an important characteristics, the *5-minute handling capacity HC*:

$$HC = \frac{300}{INT} \times P \tag{5.4}$$

HC is of great interest to architects, for designing and optimizing the specifications of the transportation system of the building.

The above is only a rough sketch of the design calculation; in practice, there are many complicating factors, refinements and variations. For instance, elevators will not have passengers going to the top floor on each trip; so on average, elevators will not travel up N floors, only a smaller number H; we can correct for this effect. Also, besides the *up-peak* traffic, other traffic situations will need to be considered. For a thorough exposition, we refer again to the classic source [5].

5.1.5 Nodes and Links

The transportation system needs to connect the sources and sinks of the traffic, throughout the building. Since typically the largest source and sink is the main lobby floor, this is the root node of the transportation network. Other major nodes can be at restaurant and cafeteria floors, at observation decks, underground floors with subway connections, *etc.*

It is often the case that there is a dominant mode of traffic that determines the overall topology of the transportation network. For ultrahigh-rise office buildings, this would be the morning incoming peak traffic, where people arriving at the main lobby want to reach their offices, at the same time.

The simplest topology would be to connect all floors directly with each other. This works for small buildings, but as the number of floors increase above about $10 - 15$, the number of stops that a fully loaded elevator makes would become too high.

Some topologies of elevator systems that evolved to cater for this traffic are shown in Fig. 5.3. The notation is: circles represent the served floors, solid lines are elevator paths, dotted lines are pedestrian paths. The grey rectangles represent the elevator groups that serve each path.

The first example is the traditional zoning topology; each elevator group serves the main lobby and a (mostly) contiguous range of floors. There is immediate connection from the entrance to all upper floors; however, floors in different zones are not connected directly. This topology limits the number of stops each elevator makes, but not their running distance. Elevators in higher zones need to spend an increasing part of their time running in the *express zone*. These elevators need to be expensive, ultrahigh-speed models. This topology is not economical for the highest buildings.

The second one, the so-called *sky lobby* system, is mostly used in ultrahigh-rise buildings. Here only the floors in the lowest zone have direct connection to the main lobby. Other zones have their own secondary lobby, which are connected to the main lobby by high-speed, two-stop shuttle elevators. People traveling to higher floors will take a shuttle, and board a local elevator from one of the secondary *sky lobbies*.

The third one is a variant of the *sky lobby* system. The sky lobby is placed high in the building, and two local elevator banks serve the middle and upper regions of the building.

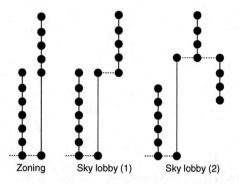

Fig. 5.3. Three typical topologies for traffic dominated by up-peak

5.2 Human–machine Interaction and Control Objectives

5.2.1 Modeling of the Traffic

In this section we briefly consider the passenger traffic of elevators, as the most representative and well-studied traffic component in buildings.

Passenger traffic is usually modeled as one or several Poisson processes; that is, on the assumption that the occurrence of each arrival event is independent, only their probability density can be given.

In a building, it is useful to classify passenger arrivals at the elevator halls into three basic modes:

Incoming: persons arriving at the building and typically going up from the main lobby; for example, employees reporting to work at office buildings in the morning.

Internal: persons moving between floors; for example, office workers attending meetings during the day.

Outgoing: persons leaving the building; for example, the evening traffic rush directed to the lobby of an office building.

As a first approximation, the traffic could be said to be composed of these modes, with their degrees varying during the day. Continuing with the example of the office building, when the *incoming* traffic dominates, we speak of an *up-peak* traffic pattern; conversely, a dominant *outgoing* traffic is called the *down-peak*. During the lunchtime period, the two modes might overlap, as some people are already coming back from lunch, while others are just leaving.

To describe the traffic quantitatively, the most useful concept is the *origin–destination (O/D) table*. This is a matrix $[o_{i,j}]$, with one row and one column for each floor. The element $o_{i,j}$ represents the relative probability that when a passenger arrival occurs, it will be from floor i to floor j. If, in addition, we have the arrival density λ, that is the expected number of arrivals in unit time, we can create a modeled arrival stream, using random number generators to generate the arrival events in exponentially distributed random intervals; and assign to each arrival an origin and destination floor, according to the relative frequencies of the O/D table.

Table 5.2. Stylized O/D table

Origin	Destination					
	1	2	3	4	5	6
1	0	10	10	10	10	10
2	1	0	1	1	1	1
3	1	1	0	1	1	1
4	1	1	1	0	1	1
5	1	1	1	1	0	1
6	1	1	1	1	1	0

As an example, a stylized O/D table for the *up-peak* traffic of a 6-floor building, where the 1st floor is the main lobby, is given in Table 5.2. We shall return to the topic of traffic modeling in the discussion of the simulation techniques later in this chapter.

5.2.2 Human–machine Interface of Elevators

The passenger-elevator interaction starts with the signaling of the travel demand, *i.e.* by registering a hall call. At present, automatic operation of elevators is usually done by push buttons, although some other interfaces (remote signals from electronic tags, reservation of service from local-area networks, *etc.*) are also being developed. For feedback to the passengers, the elevator system gives visual signals by lamps, audible signals by chimes and gongs, in addition to observing directly when a car arrives and the door opens.

In the case of group control, the passenger is in fact interacting with the group controller, not directly with the cars. This allows the overall optimization of the operation, since passengers would almost certainly use "greedy" strategies if they could directly control the car movements. That is, for overall efficiency, the group controller often makes cars bypass a hall call on the way to other calls; if the passenger waiting at that floor could stop the car, he probably would. Also, cars stop at some intermediate floors on the way to a car call; if the passenger in the car had full control, he would probably go directly to his destination without stopping.

In the future, we can assume at least the following developments:

- Personal service (passive and active identification for access control, tracking, *etc.*)
- Improved control interfaces (mobile phone, voice, gesture, remote, *etc.*)

- Multimedia information (broadcasts, personalized messages, advertisement etc)
- Improved special services (for handicapped and senior persons)
- Improved emergency and security supervision (for evacuation, *etc.*)

These require extensive further research in sensing, actuators, and intelligence.

5.2.3 Human–machine Interface of Escalators and Other Equipment

Compared with the elevator, some other building transportation equipment employ much less sophisticated interfaces. Escalators usually show only their operating direction by lighted arrows. Sometimes a notice is displayed when they are in an energy-saving operating mode, slowing or stopping without traffic and restarting on sensing a passenger.

Dumbwaiters and freight elevators are assumed to be operated not by the general public but by trained personnel. This allows the use of more custom-made controls.

5.2.4 Control Objectives

In general, the objective of a transportation system is to support the traffic flow with the minimum of delay and cost. We should be aware that these are conflicting requirements: with more resources the traffic handling becomes better, but at a higher cost.

In the case of transportation systems in buildings, the primary goal becomes different for passenger and cargo system. For passengers, we have to provide acceptable service not only on the average, but to each and every individual. On the other hand, cargo transfers can be scheduled more tolerantly, and the primary objective becomes achieving the maximum throughput with minimal equipment and running costs.

We list here some recommended performance requirements for passenger elevators from [2]:

- **Waiting time [s]** 30 (commercial) – 60 (residential)
- **Journey time [s]** 60 – 90
- **Handling capacity [% of population in 5 min]** 10 – 20

The actual requirements depend on many factors: the occupation and use of the building, the expected traffic patterns, *etc.*, and cannot be formulated easily as ready-made rules. Since poor service can cripple a building, it is very important to rely on professional expertise in deciding the requirement levels. Most elevator companies, as well as independent elevator consultants, offer consulting services for those wishing to plan transportation facilities for buildings. These services are supported by their accumulated experience, and also by increasingly sophisticated software tools; see *e.g.* [12, 13].

6

Queuing Models

In this chapter we will consider some quantitative properties of the transportation systems. The models that are introduced here are "macroscopic": we deliberately forget about the details of the individual components and their operation, in order to find some laws that apply to their overall, statistical behavior. Thus these models cannot be used for control; but they do give some insights that can guide the development of control techniques.

6.1 General Overview of Queuing Models

A well-studied model of transportation systems is the queuing model. It is based on the notion of dealing only with a subset of the discrete events that occur in the system, and trying to derive the statistical properties for the temporal sequence of these events. For instance, in a passenger transportation system, such "important" events might be the arrival of a passenger, and the start and end of a journey. The duration of the intervals between such events is assumed to be random, governed by some probability distribution.

In the queuing approach, it is useful to consider *arrivals* as a process; we talk of passengers or parcels being *generated* by the *arrival process*. Similarly, their actual travel is another *service process*. In general, there can be several arrival processes, which can be connected to a set of service processes, according to some specific *service discipline*, such as *first in, first out*.

We arrive at the most elementary queuing model by assuming a single process of independent arrivals, and similarly a single service process. Thus we merge together all service events happening in the building, as if they were taking place in a single virtual queue.

From the condition of independence, it is possible to derive the distribution $f(t)$ of the interarrival times t, which happens to be exponential:

$$f(t) = \alpha e^{-\alpha t}$$

If we also assume that the service intervals are exponentially distributed, the resulting model is called the $M/M/1$ model, in the Erlang notation. We can generalize it to allow random but not exponential service times, in the most general case arriving finally at the $M/G/1$ model. We cannot deal here with any of the further generalizations, e.g. to multiple servers, different queuing disciplines etc.

At first, it is not clear why should we assume a random service process; after all, elevators, AGVs, or stacking cranes are operating strictly deterministically, running and stopping exactly as commanded by their controller electronics.

However, although individual vehicles do in fact behave deterministically, they interact with their load (passengers or packages), which is where the random (stochastic) behavior enters the system. Therefore, in reality the overall system behavior is stochastic, and the completely deterministic model of Section 5.1.1 is only useful as a first approximation.

As an illustrative example, we shall explore the elevator system as a queuing system in some more detail below.

6.2 Queuing Models for Elevator Systems

From the point of view of elevator control, each passenger has at least the following random characteristics, with the first two being the most important:

1. the *origin* and *destination* floors,
2. the arrival time at the *origin floor*,
3. boarding and departing delay times,
4. the probability of *balking*, i.e. of leaving the waiting queue if the waiting time is estimated to be too long, or refraining from boarding a car if it is too crowded.

The resulting random process has been considered by many researchers. It seems natural to view the interaction of passengers and elevators as a queuing process, as described below.

At a given floor and direction, passengers will face one of the following situations. They will either find a car available with doors open, or the first passenger registers a hall call, and subsequent passengers queue up until a car comes. All waiting passengers, up to the capacity of the car, are then served together, according to the *bulk service* queuing discipline [14].

6.2.1 The Simplest Case: M/M/1 Model

As a first approximation, we can proceed by disregarding the immediate effect of passengers on car movements, and assume that the two processes are independent. This model is just a superficial, phenomenological description of the passenger and car arrivals at a given elevator hall. In this model, passengers arrive randomly, and board a car when available. Cars also move randomly, and take on passengers if there are any of them waiting. This is an approximation, of course: in reality, cars do not move independently of passenger arrivals, but rather they respond to calls. The two processes are illustrated in Fig. 6.1.

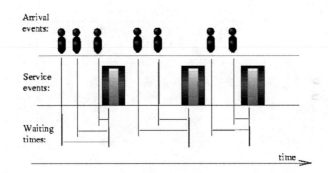

Fig. 6.1. A simple queuing model of the elevator system

Mathematically it is especially easy to treat the case when we assume that

- the passenger arrival is a Poisson process, with exponential interarrival times, and an arrival rate λ;
- the elevator service is also a Poisson process, with a service rate μ.

It is convenient to define the "utilization ratio" of the system $\rho = \lambda/\mu$. Obviously, we need $\rho < 1$ for the system to be viable; otherwise in the long run there would be more and more passengers queuing up who never receive service.

Although such a simple model cannot describe the actual elevator service process, we can still gain some insight from it, if we consider the following result for the M/M/1 model from elementary queuing theory:

The waiting time w is given as a function of ρ by

$$w = \frac{1}{\lambda}\frac{\rho}{1-\rho} \tag{6.1}$$

that is, it diverges as ρ approaches unity.

As an example of application, let us consider the question of estimating the effect of improvements in an elevator service process, on the system performance.

Suppose, for instance, that we have a system with an hourly passenger arrival rate $\lambda = 900$, and our elevators have a service rate of $\mu = 1000$ passengers/h. Now what would be the improvement if we could increase the service rate by 5%, to $\mu' = 1050$?

Our first guess might be, "not much". However, this is a mistake. Our system is operating near capacity, and a queuing system with $\rho \approx 1$ has a very nonlinear response curve. Let us plot it for our system, before and after the 5% improvement (see Fig. 6.2).

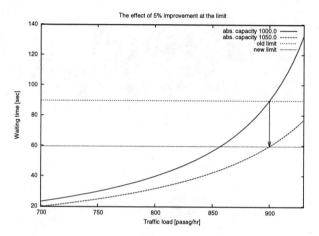

Fig. 6.2. The effect of improved service on the performance in the M/M/1 model

We see that even with such a small improvement in the service rate, we get an enormous improvement in the performance. This kind of observation has implications to the decisions we make, when directing the research effort in transportation system control.

6.2.2 A More General Model: M/G/1

If the passenger arrival process and the car service process can be assumed independent, the relatively simple model of Bayley [14] can be used. This model was applied to elevators by Alexandris, Barney, and Harris [15–17], and also studied by Hirasawa [18, 19] and developed further by others [20]. A few important results are listed below. If the service time probability distribution is approximated by a k-Erlang distribution with mean M, the probability distribution of the waiting time τ is given by

$$p(\tau) = \frac{1}{M} \int_\tau^\infty \frac{(\mu\tau)^{k-1}e^{-\mu t}}{(k-1)!} \mu dt = \frac{1}{M}e^{-\frac{k}{M}\tau} \sum_{n=0}^{k-1} \frac{\left(\frac{k}{M}\tau\right)^n}{n!} \tag{6.2}$$

so the average waiting time is given by

$$W = \int_0^\infty tp(t)\mathrm{d}t = \frac{M}{2}\frac{k+1}{k} \tag{6.3}$$

Here k is the parameter of the Erlang distribution family, ranging from $k = 1$ (corresponding to an exponential distribution) to $k \to \infty$ (the Gaussian limit). Therefore, for $k \to \infty$ (deterministic service, all service times equal), from Equation (6.3) we get $W \to \frac{M}{2}$, which is the lowest possible value. This is in fact the assumption behind the usual calculation of the expected waiting time as $\frac{INTV}{2}$ in the conventional design calculation of Equation (5.4).

For $k = 1$ (exponential service), we get $W = M$, which is the well-known *taxicab paradox*: an observer arriving at any instant between service times always has to wait on average for the full service interval. (This counterintuitive result can be informally explained by considering that the observer is more likely to experience a service event with long waiting interval than a short one.) It is also easy to verify that the probability of a long wait (waiting time over some given limit a, usually taken as $a = 60$ s) is given by

$$W_a = 1 - \int_0^a p(t)dt = 1 - \frac{1}{M}\sum_{n=0}^{k-1}\frac{\Gamma\left(1+n, 0, \frac{k}{M}a\right)}{\left(\frac{k}{M}\right)^n n!} \tag{6.4}$$

which also increases for low values of k. These results, especially the rapidly increasing occurrence rate of long waiting times, motivated several approaches to the group control.

The M/G/1 model can be used to get a qualitative understanding of the effect of the service-time distribution on the system performance. We can say that as we improve control, we can steer the system away from the Poisson limit (complete randomness) toward the deterministic limit (equal service intervals). We can also tell how much improvement this can buy us: at an extreme, we can gain $2 : 1$ better performance "just" by better control, that is, with no change whatsoever in the physical system resources. This is such a substantial potential for improvement that it completely justifies the efforts spent on the development of better control methods.

As a sidenote, we can also mention that on the basis of queuing models, several authors have proposed improved traffic analysis calculations [21]. For example, Peters [22] assumes deterministic service times (equal service intervals), and calculates S and RTT.

Modeling Techniques for Discrete Event Systems

The transportation systems in a building can be modeled on a computer on several levels. We shall not be concerned here with models describing the detailed physical behavior of system components, such as the feedback control of the drives, or the vibration of cages. Such models are often used in the design of the components, but they are unnecessarily costly in terms of human and computational resources for our purposes.

We shall deal here only with models that describe the systems as a set of interconnected components, each with a predefined set of state variables, which can take on only discrete values. We call the transition of a state variable from one value to another an "event". The discipline of discrete event simulation deals with the construction and evaluation of systems, modeled with a temporal sequence of events.

For discrete event simulations, there are many available tools and software packages, both commercial [23] and free. Some examples are found in Banks *et al.* [24]. There is also simulation software dedicated to the modeling of elevator systems [13].

As a preliminary to the discussion of simulation techniques, in Section 7.1 we first describe briefly how field studies are used to help in building, verifying and tuning the simulation models.

Next, in Sections 7.2 and 7.3 we shall review some basic background from the literature, then introduce the specific methodology that we have been using for transportation research.

Finally, in Section 7.4 we review an application example; based on traffic data obtained from field studies, we construct a simplified model of the passenger circulation in an existing commercial building.

7.1 Field Studies

The existing analytical methods for the study of elevator group control systems have limited value, mainly because they do not account for the details of control, and therefore cannot distinguish between systems with different control algorithms. This makes analytical methods insufficient if the goal is the development and comparison of group control methods.

Because of this, at present the two main methods of obtaining detailed data about elevator groups are experimental:

- traffic data logging, and
- Monte-Carlo modeling (simulation).

Modern group controllers are usually provided with *data logging* capability. If the most important events, like car movements, call registration and service, are recorded together with the time of occurrence, the logged data can be analyzed for many purposes. An example of the elevator travel diagram is shown in Fig. 7.1. Statistical analysis can indicate the traffic and service distributions, both long-term averages and fluctuations [25]. It is also possible to search for special events, like *e.g.* typical situations for applying control heuristics. However, since individual passenger arrivals and services are not observable, the traffic flow is not measurable directly. The number of passengers who travel between different floors can only be inferred from the observable data, using a suitable statistical procedure.

There are several proposals in the elevator literature dealing with the problem of finding the underlying hidden events, using such observable data as the number of registered car calls at each stop, the temporal change of the car load during leaving and boarding, the signal from the photocells of the door safety mechanism, *etc.* The *Inverse S/P Method* of Al-Sharif and Barney [26, 27] uses the relationship between the passenger load P and the number of discharging stops S to estimate the passenger counts in each car:

$$S = N \left[1 - \left(\frac{N-1}{N} \right)^P \right]. \tag{7.1}$$

This equation was introduced in Section 5.1.4, cf. Equation (5.1). By inverting the relation, we can obtain an estimate of the expected number of passengers as

$$E(P) = \frac{\ln \left(\frac{N-S}{N} \right)}{\ln \left(\frac{N-1}{N} \right)} \tag{7.2}$$

from the observed values of stops S for each trip of a car. Here, as in Equation (5.1), equal probabilities are assumed for stops at each floor, and we assume that no prior knowledge is available about the distribution of P. In [27], the application and further extensions of Equation (7.2) are discussed in detail.

The operation of the elevator group can also be studied by *simulation*. For the same elevator system as in Fig. 7.1, an example of the elevator travel diagram for a simulation is shown in Fig. 7.2. Since the elevator dynamics is deterministic, and the control algorithm is known, the only source of error in the simulation is introduced by the need to reproduce traffic situations in buildings. Otherwise, simulation can provide performance samples for any elevator system, that can be statistically analyzed, *e.g.* to evaluate group-control methods.

In the following chapters, simulation programs are used extensively in the development and evaluation of new group-control methods. For traffic parameters, the data was obtained from actual operation, and the *inverse S/P method* was used to find initial traffic-density estimates. These were then corrected by running simulation experiments and comparing the resulting waiting-time distributions with the logged waiting times, then adjusting the traffic densities until a reasonable agreement was obtained.

In the next section we discuss in some detail the discrete event simulation techniques that are used in Monte-Carlo modeling of transportation systems.

7.2 Monte-Carlo Modeling

A transportation system can be approximated at a "mesoscopic" level, *i.e.* at the level of medium-scale components and actors: elevator cars, passengers, AGVs, group controllers, *etc.* These constituting elements are assumed to be atomic, and are governed by predefined behaviorial laws, with deterministic and stochastic components.

This description enables us to apply a simulation-based methodology for analyzing the system behavior In a simulation study, the system is modeled by a computer program, and its behaviors observed as the temporal evolution of the model.

This methodology is widely used not only in transportation systems, but in the study of all kinds of OR (operations research) problems that involve dynamics, randomness, and logistics.

7.2.1 Simulation Techniques

Following Mitrani [28], we will review briefly the main steps needed in simulation studies:

1. Construction of the model
2. Design of the simulation experiment
3. Implementation of the simulation program
4. Execution of the simulations
5. Analysis of the simulation results

7.2.2 Modeling by ESM-based Methodology

Simulation models need a conceptual framework before any actual coding can start. As a modeling framework, the *extended state machine* (ESM) introduced by Ostroff [29] strikes an appropriate balance between simplicity and generality.

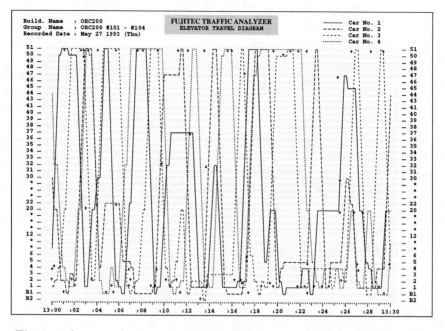

Fig. 7.1. An example of an elevator travel diagram plotted from logged traffic

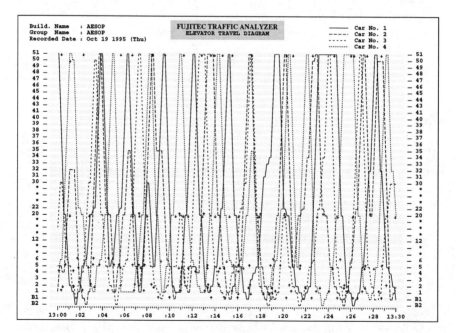

Fig. 7.2. Elevator travel diagram for the same building from simulation

The original formulation is derived from an even simpler framework, the CSP (communicating sequential processes) of Hoare [30]. In a somewhat simplified point of view, a CSP model of a concurrent system is composed of independent processes, each of which executes its own sequence of tasks, stepping from state to state. There is also a set of "channels" that are used for communication. Channels have two endpoints, for sending and receiving a message. Communication events over channels are the exception to independence, by allowing two participating processes to synchronize their execution. Communication takes place when both processes connected to the ends of a channel become ready: one to send a message and the other to receive it.

Although the CSP model was introduced chiefly as a theoretical formulation, allowing the formal analysis of processes to derive such properties as the presence or absence of deadlocks, it has also been very influential in computer applications. For instance, the design of the computer language OCCAM [31] used for the "Transputer" microprocessor [32] was based on CSP.

The chief difference between CSP and the ESM model is the introduction of "data variables" in the latter. In the view of a CSP model, the state of the "world" (the system under investigation) is completely described by the set of states of the processes comprising the model. This is very clean and simple, but it makes modeling of realistic systems unwieldy.

As an example, suppose that we have an elevator that serves 20 floors. To describe its position as the states of a CSP process, this process needs at least 20 states; so far not too bad. But suppose further that we want to model the passengers waiting on each floor. Now the problem starts getting out of hand: if there can be $0, 1, \ldots 50$ passengers on each floor, we need 20 more processes, each with 50 states. Also, we need to set up the appropriate channels to provide triggers for state transitions in each process. Although we can imagine setting up such a model by some automated process, the result would be a gigantic, cluttered diagram with hardly any descriptive power.

In the ESM formulation, we allow state variables that are not directly represented as process states. This makes the model much more compact: we will still make explicit those states that are important for describing the temporal evolution, but we can keep the rest of the variables implicit. Thus, if we draw a state transition diagram it will accurately reflect the structure of the model.

7.3 The ESM Framework for Simulations

7.3.1 The ESM Model for Discrete Event Simulation

Elevators or AGVs can be treated as discrete event systems, but with strong algorithmic components.

Though many methods have been proposed to describe discrete event systems formally, and there are many tools to simulate them, these methods are usually not used for modeling elevators. Because of the complex algorithms, elevator simulations were traditionally done in procedural programming languages, like Fortran, C, *etc.*

Here we adopted a variant of the ESM model, for the following reasons:

- In the systems that we want to model, there are many states and state transitions. If we were to use a procedural programming language for modeling, it would be difficult for us to develop, debug and maintain.
- Conversely, pure discrete event formalisms, like *e.g.* the Petri net, are not well suited for describing the algorithmic aspects, such as the group-control programs of elevators, or the scheduling of AGVs.

We have therefore developed a combined formalism, based on the ESM of [29], which allows the transparent modeling of the state-transition structure, and at the same time provides a correct description of the control algorithms.

We will define the ESM model below.

Definition of the ESM

Definition 7.1 An ESM M is defined as a 7-tuple $\{S, L, P, M, E, A, T\}$

- *S: the set of state parameters determining the unique system state*
- *L: the set of state labels for identifying a subset of S*
- *P: the set of message ports, constructed as the union of sending ports P^s and receiving ports P^r: $P = P^s \cup P^r$*
- *M: the set of messages exchanged between ESMs*
- *E: the set of events that trigger the transition of states*
- *A: the set of actions executed in the transition of states*
- *T: the set of paths for transitions*

Actions and Events

The set of actions A is divided into two subsets.

- $A^{send} := P^s \times M$ the set of actions of sending messages
- $A^{calc} := \{f_i : L \times E \times S \rightarrow S\}$ the set of actions updating the internal state parameters.

The set of events E is defined as $E = E^{rec}$ with E^{delay}, where

- $E^{delay} := \{delay(t)|t \in \mathbf{R}\}$ the set of timer events,
- $E^{rec} := P^r \times M$ the set of events of receiving messages.

The two types of events: the timer expiring and the receiving of messages, are described graphically in Fig. 7.3.

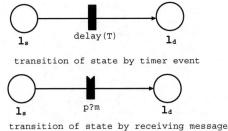

transition of state by timer event

transition of state by receiving message

Fig. 7.3. Graphical notations of events

State Transition Paths and Branching

When an event triggers the state transition of an ESM, it will go through one of the transition paths T into the next state. The actual path traversed can depend on the internal states S; this is how we make the model deterministic, and here is the connection between the algorithmic and discrete-event aspects.

We define T by introducing branching and action sequences.

To branch conditionally depending on the internal states of the system, we define the set of branching functions B as the set of predicates $B : S \rightarrow \{true, false\}$

Now we can define the action sequence D as

$$D ::= L | A^{send}, D | A^{calc}, D | B, D^T, D^F$$

Using these, the set of transitions is defined as $T ::= LED$

If we set $l_s \in L$, $e \in E$, and $a = a_1, b_1, l_d^t, l_d^f$ ($a_1 \in A^{send}, b_1 \in B, l_d^t, l_d^f \in L$), the transition $t = \{l_s, e, a\} \in T$ means "When the system label is l_s, after triggering by the event e, with the transition of system states we execute the action sequence a. "

The sending of messages, updating the internal states of system, branching by predicates after the occurrence of events are described graphically by symbols in Fig. 7.4.

7.3.2 Communication Between ESMs

For constructing the simulation model for large-scale buildings including elevators, AGVs, and traffic flow such as goods or persons, we have added the concept of *storages* to the ESM model, to handle the dynamical topology of communicating channels in the model.

Definition of the *Storage*

A *storage* has the following three characteristics:

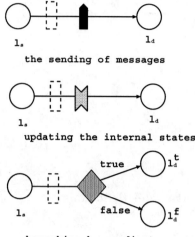

the sending of messages

updating the internal states

branching by predicate

Fig. 7.4. Graphical notations of actions

- It provides the generation and the disappearance of ESMs.
- It can control the dynamical communication channel between ESMs.
- It can control the location of the ESM in the system.

Definition 7.2 *A storage S is defined by a 9-tuple $\{\Theta, L, I, P^{out}, P^{in}, C_{in}, F, A, O\}$.*

- *Θ: the set of ESMs or (other) storages.*
- *L: the set of labels.*
- *I: the set of indexes.*
- *P^{out}: the set of outer ports.*
- *P^{in}: the set of inner ports.*
- *C_{in}: the set of inner channels.*
- *$F = F^{out} \cup F^{in}$: the set of functions switching the inner channel connections.*
  - *$F^{out} : M \times I \times L \rightarrow P^{out}$*
  - *$F^{in} : P^{out} \times M \times L \times I \rightarrow P^{in}$*
- *$A := \{enter, step, exit\}$: the set of actions.*
- *$O := O^s \cup O^r$ the set of ports for the passage of ESMs, called an object port.*

The communication between ESMs, as mediated by storages, is defined as follows.

Definition 7.3 *Suppose two storages*
$S_i = \{\Theta_i, L_i, I_i, P_i^{out}, P_i^{in}, C_i^{in}, F_i\}(i = 1, 2)$
each contain an ESM M_1, M_2, respectively; then the one-way message channel

between M_1 and M_2 is represented as

$$C = \{(p_1^{out}, f_1^{out}), (p_2^{in}, f_2^{in}) | p_1^{out} \in P^{out}, f_1^{out} \in F^{out}, p_2^{in} \in P^{in}, f_2^{in} \in F^{in}\}$$

Sending Messages and Objects Between Storages

The procedure of sending a message m from M_1 to M_2 is described in Table 7.1.

Table 7.1. The order of dynamical communication

M_1	S_1	S_2	M_2
$p_1!m$			
	$F^{out}!(l, i, m)$		
		$F^{in}?(l, i, m)$	
			$p_2?m$

The movement of ESMs between storages is done by using the *object channel* defined below.

Definition 7.4 *The object channel between two ESMs*

$$S_i = \{\Theta_i, L_i, I_i, P_i^{out}, P_i^{in}, C_i^{in}, F_i\}(i = 1, 2)$$

is defined by

$$C_o = \{(s_1, o_1'), (s_2, o_2') | o_1' \in O_1, o_2' \in O_2\}$$

7.3.3 Tools for Defining the ESM Model

In our research, we have developed an integrated GUI tool for handling the total development process, from constructing the ESM model to the execution of the simulation. We have implemented both the ESMs and the storages in the object oriented language Java. The ESMs and the storages are both described as the objects, and messages are implemented as method calls, without any buffering overhead.

The following is the procedure of constructing a model of simulation:

1. Define the ESMs by GUI tools.
2. Define the inter-ESMs communications.
3. Generate automatically the simulation code in an intermediate language [33].
4. Define the algorithmic parts directly in the Java language.
5. The Java language translated from the intermediate language generated in Step 3 and the handwritten Java code of Step 4 are merged into the execution code.

7.3.4 Implementation of the Simulation Program

There are some basic components used by all simulation systems:

1. Discrete event scheduler
2. Random number generator
3. Statistical processor

Discrete Event Scheduler

In general, the discrete event scheduler of the simulation system would need to implement a priority queue, which supports at least the following operations:

- Inserting an event in the queue.
- Checking whether the next event has a due time earlier than the termination time of the simulation.
- Accessing and removing the event (or one of the events) with the earliest due time.
- Accessing and removing an event that had been superseded by another related event.

As the simulation progresses, at state transitions new events are generated and registered in the event queue. Also, some existing events might need to be removed from the queue.

In the following, we discuss briefly only the case of the ESM methodology. It is in a sense special, in that each ESM (state machine) has exactly one timer. Thus, for each ESM we can permanently assign an event, which represents the next scheduled firing time for the timer, and update its value as needed. This updating automatically takes care of the removal of the undesired old event, which would be required in some other methodologies.

The simplest implementation of the event queue might use an array, which would be scanned at each time step, and the earliest event executed. If the array is not sorted, this operation has a computational cost linear in the number of ESMs, which in the case of small or medium-size systems could be perfectly acceptable.

A more efficient implementation might use a *heap*, that is, a tree structure with logarithmic insertion cost, and constant cost for accessing the earliest event. Sophisticated simulation systems make considerable efforts to optimize the scheduler, which can significantly affect the overall efficiency.

Random-Number Generator

In simulation studies, by *random numbers* we usually mean pseudorandom numbers, generated by a suitable algorithm. Much research has been done on random-number generators, since it has been found that poor-quality random sequences can distort the results of simulation experiments. The interested reader should consult the very extensive literature; however, we note that we are using the random-number generators found in the standard library of the C and Java languages.

Statistical Processing

The simulation experiment results in a record of a sample evolution of the system, that is, a list of the time points of state transitions. Usually we are interested in some specific events, or rather, in event pairs: the time interval between two related events, like the arrival and service instants of each passenger. Therefore we will record a sample set of such intervals, and process it to obtain a statistical estimate of some value of interest.

The most common case is when we want to estimate the expectation of a system performance value, such as the passenger waiting times, or the total times in the system (which includes the waiting times and the service times). The sample average is the maximum likelihood estimator of the expectation, so a simple averaging will get the required value.

However, we must be careful to check for some possible stumbling points. The first thing to do is a "sanity check": are the results consistent with the initial assumptions, or not? For instance, obvious outliers in the sample distribution might be caused by mistakes in the model, or even by bugs in the simulation program. Also, it is useful to try to compare visually the simulation run and the collected samples; unexpected trends or correlations can also signify modeling problems.

Next, we need to look at the sample variance: if it is too large, the simulation results cannot be trusted. To be more precise: if we want to compare two simulation experiments, we would need to perform a significance test on the results, using *e.g.* the sample means and sample variances, to be able to say whether there is a significant difference between them or not.

Furthermore, simulation experiments are usually performed on the assumption that the system is in a steady state of operation. However, if a transportation system is overloaded, it cannot reach a steady state. Instead, queues will grow, and the numerical results will depend on how long simulation intervals had been used. In such cases, the numerical results are meaningless in themselves, although they might still be used in qualitative comparisons, *e.g.* of different control strategies.

Again we have to refer to the simulation textbooks for more detailed advice on the design of the statistical postprocessing of the simulation results.

In the next section, we use a specific study as an example, to introduce modeling by the ESM framework.

7.4 Modeling Cooperating Elevators and AGVs by the ESM Methodology

In this section, as an example of the methodology, we introduce a modeling study on a proposal of providing an alternative solution to goods transport in buildings. Here we describe the observed data that we used to build the models, and present a simplified simulation model. On the basis of this model, we have studied the design of a new control strategy, which will be described later in Chapter 17.

For transportation, traditional architectural design relies mostly on elevators, which are well fitted for passenger service, but less well to the delivery of packages and other materials. Transporting goods by people riding elevators is not necessarily the best solution, for several reasons:

- Not only the goods, but the accompanying persons also will be moved, thus reducing the available handling capacity, and incurring personnel costs;
- Goods don't need the same quality, space, and speed of service as people, thus elevators are overdesigned for them;
- Point-to-point, just-in-time package delivery is becoming common in urban life, so it is going to be expected inside buildings, too. This would require automatic package handling, instead of the present manual one.

There are proposals for adapting various automatic material handling systems, like AGVs (automated guided vehicles) or linear-motor-based systems, to provide vertical transportation for goods in buildings [34]. Also, besides traditional elevators, there is growing interest in linear-motor elevators and other new transportation technology. These, and their combination with elevators, could revolutionize the vertical transportation design of buildings.

Another recent trend is the growing interest in the new discipline of complex systems theory. This field seems promising for the efficient control of distributed transportation systems, especially those like AGVs that have dynamically changing configuration,

However, at present much basic research is still to be done, before we can evaluate the merits and practical realizability of such systems.

In order to contribute to the systematic revision of the vertical transportation systems of large buildings, we have started to study the following basic topics:

1. Measurement and analysis of the actual traffic in existing buildings;
2. Development of a simulation methodology and models for present and proposed transportation systems;
3. Research on the distributed autonomous control of material handling systems in buildings.

The present section reports on our progress in the above research topics.

7.4.1 Traffic Survey as the Starting Point for Simulations

The knowledge of the actual traffic situation in buildings is a necessary condition for improving the transportation system. Still, it seems that there are not many reports of quantitative data about passenger and goods traffic in buildings. We have therefore conducted a traffic survey, with the main goal of examining the ratio of persons bringing packages with them. We report the results for a building with the specifications of Table 7.2.

Table 7.2. Specification of the target building

Use of building:	Office building
Style of use:	Single occupant
Number of floors:	15 floors
Number of persons:	About 1300 persons
Number of cars:	9 cars (6+1 passenger lifts, 2 freight lifts)

We have posted a number of observers at some of the elevator halls in the target building, who counted visually the passengers and their packages. In our survey we have counted only things like documents, magazines, newspapers, lunchboxes *etc.* as luggage, without including briefcases *etc.* that are normally carried by people commuting to work. The traffic survey was performed for the period of reporting for work, for lunchtime, for the time of leaving work, and for the intervening offpeak periods. We report the data for survey in Table 7.3.

Additionally, in this building some packages and mail for each office are delivered by a central delivery service, using carts. Table 7.3 does not contain the data of this service. From interviewing the administrator, we know that the rate of packages handled by the delivery service is about 18% of the total.

We can draw some conclusions from the results.

The results of Table 7.3 show that the rate of people with packages is the lowest for the morning peak period. Generally speaking, the number of elevators is determined by considering the handling capacity in the morning peak time, when usually the traffic is the most concentrated. In that period, using alternative methods for package handling would not reduce the passenger traffic too much.

However, there are buildings where the critical traffic peak is at times other than the incoming peak. At such sites, we can expect that substituting other devices for package handling could contribute to a better handling capacity and allow a reduction of the number of lifts.

Table 7.3. The number of passengers and the rate of people with packages

Place	Time	Rate of packages [%]	Number of passengers
1st floor	Morning peak	2	1018
	Offpeak time (a.m.)	28	324
	Lunch time	18	288
	Offpeak time (p.m.)	61	382
	Quitting time	45	223
Locker-room floor	Morning peak	12	774
Restaurant floor	Lunch time	11	1386

7.4.2 A Simplified Model of the Traffic in the Building

If we define port labels, object ports (sending, receiving), and message ports by symbols as in Fig 7.5, we can define the storage corresponding to the elevator hall as in Fig. 7.6.

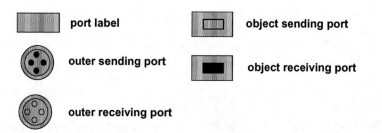

Fig. 7.5. Graphical definition for elements of the storage

Furthermore, the definition of the total elevator model is as in Fig. 7.7.

The above model is at present not a complete one, and there is need for further research, using observations from several buildings. However, this model was found to be sufficiently correct to allow the development of a new cooperative control method between elevators and AGVs, which is described in Chapter 17.

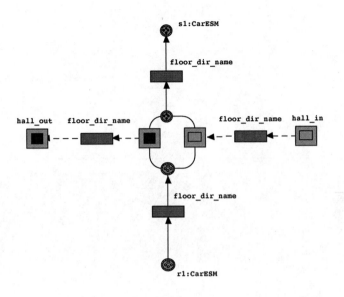

Fig. 7.6. Figures for storage of elevator hall

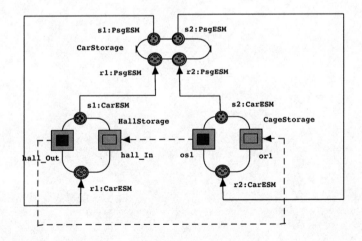

Fig. 7.7. Total definition for elevators

8

Scheduling Models with Transportation

Introduction

Everyone knows what scheduling means, and it is done in everyday life by all of us. However, it may not be realized yet by those other than specialists in the field that the scheduling tends to be a critical business process in contemporary industries where the stagnating economy under global competition requires efficient small-lot production for customer satisfaction, and hence it has been a main target of research in management, system and computation sciences, among others. Thus, it is expected that the numbers involved in scheduling research, its application and practice will further increase in the near future; this introduction is intended as a short guide from the view point of combinatorial optimization.

We start this chapter with the job shop scheduling model that does not consider transportation, because many transportation scheduling models are based on the job shop models. Then, we discuss shortly some automated job-shop and robotic cell models in factories. In Chapter 20 we shall see as an extension the case of a S/R (storage and retrieval) machine model in automated warehouses, as transportation scheduling models.

8.1 Jobshop Scheduling Problems

Scheduling is, simply speaking, the allocation of limited resources to jobs over time. Resources may be machine tools, operators and/or others. A job (also referred to as an order, lot, and other terminologies) usually consists of one or more operations or tasks to be executed on resources to satisfy a specified demand (*e.g.* making a product or a part). If more than one job has to be executed on a (scarce) resource sequentially, a processing sequence of the jobs will need to be decided. If there is more than one resource available to process a job, an allocation of the resources to it is needed. These decisions may result in a schedule, *i.e.* times when every job starts and finishes on each resource, under the policy that every job should finish as early as possible (it may not be the JIT (just-in-time) one). There may exist many constraints for schedules to be feasible in the real-world. At worst, it can happen that no feasible schedule exists. However, if feasible schedules can be found, one or more optimal schedules need to be selected for the system to run efficiently, or for other objectives. The job-shop scheduling problem that we consider here is a typical model of such scheduling problems.

The fundamental characteristics of the job-shop scheduling problem (JSP) is defined as follows.

JSP: Let us have n kinds of jobs to be processed on m machines, satisfying the constraints listed below. What we call here *machines* are not only actual machine tools in factories; they can also be CPUs of computers, trucks of a transportation network, or many other things. In short, they represent resources that are most likely to become bottlenecks. Then the goal is to find a feasible schedule for the jobs on each machine such that a given objective function is optimized. It is more realistic to consider a multiobjective formulation; however, it is beyond our scope (see,*e.g.* [35] and [36] for multiobjective scheduling).

Constraints:

1. For each job with multiple operations, the required sequence of the operations on each job,*i.e.* the sequence of processes the job goes through is given and fixed due to the technological constraint (such a sequence is called processing routing); the routing can be different for each job.
2. Each process the jobs go through has only one machine.
3. At any instant, each job is processed on at most one machine, and each machine processes at most one job. Also, once the processing of a job has started, it cannot be interrupted until it is finished. The pre-emption or splitting of processing may be allowed in some cases, especially in computer systems, but we here focus on the nonpre-emption (see, e.g. [37] for pre-emptive scheduling).
4. All jobs to be scheduled are available at the start,*i.e.* at time zero.
5. Processing time for each job on each machine (*i.e.* each operation) is deterministic and given in advance.
6. There may or may not be be buffers between the machines, where jobs can wait temporarily.

The job-shop consisting of the above fundamental characteristics, is called the "static" model. If the possible earliest start (or release) time is specified for each job, it is a "dynamic" model, and if processing times are random, a "stochastic" model. We focus on the static model, unless stated otherwise (see *e.g.* [38] for the dynamic and stochastic models). By relaxing or restricting some conditions of the JSP, we can generate various subclasses of the JSP for which optimal scheduling methods may be quite different. Representative examples are:

- One-machine shop problem (1SP): when there is only a single machine to process the jobs;
- Parallel shop problem (PSP): when there are multiple machines, but each job is processed only once on a single machine;
- Flowshop problem (FSP): when the processing routing is identical for each job, which thus can be said to be flowing through the system;
- Flexible flowshop problem (FFSP): a flow-shop with one or more processes consisting of multiple machines;
- Open shop problem (OSP): when the routing is not given, but they are part of the decision variables;

and many others.

To better understand the above, we first consider the 2-machine FSP with unlimited capacity of buffer between machines. Figure 8.1 shows a schematic block diagram. FSP that can be regarded as a model of manufacturing systems; while OSP and PSP can be seen as models of communication systems or computer systems.

Fig. 8.1. A 2-machine flowshop

Table 8.1. An example problem with 5 jobs

Job	1	2	3	4	5
Processing time on M_A	4	3	4	2	1
Processing time on M_B	1	2	3	3	3

Assume $\pi = (1, 2, \cdots, n)$ to be a processing order of the n jobs without loss of generality (otherwise, renumber jobs), and let $p_A(j)$ and $p_B(j)$ be the processing times of job j on two machines A and B, respectively, then starting (and finishing) time of the jth job, $s_A(j)$ ($f_A(j)$, resp.) on M_A and $s_B(j)$ ($f_B(j)$, resp.) on M_B are, respectively, given by

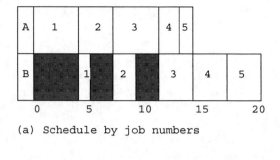

(a) Schedule by job numbers

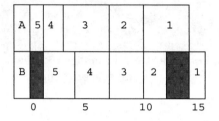

(b) Schedule optimized for completion time

Fig. 8.2. Schedules for the 2-machine flowshop

$$s_A(j) = f_A(j-1),$$
$$f_A(j) = f_A(j-1) + p_A(j)$$
$$s_B(j) = f_B(j-1) + \max\{f_A(j) - f_B(j-1), 0\}$$
$$f_B(j) = f_B(j-1) + \max\{f_A(j) - f_B(j-1), 0\} + p_B(j)$$
$$= f_B(j-1) + I_j + p_B(j), \quad j = 1, 2, \cdots, n \qquad (8.1)$$

where $f_A(0) = 0$. This formulation implies that when a job is finished on machine M_A, it will immediately get out to the buffer, and then the processing of a new job, if it exists, is started. Thus, machine M_A is always busy until it has finished all jobs. On the other hand, machine M_B is idle (or starving) after finishing a job, if there are no jobs in the buffer and the next job is still processed on machine M_A (*i.e.* if $I_j = f_A(j) - f_B(j-1) > 0$). The makespan $F_{\max}(\pi) \equiv f_B(n)$, obviously, is minimized by minimizing the idle time on machine M_B. That is,

$$F_{\max}(\pi) = I(\pi) + P \qquad (8.2)$$

where P is the total sum of processing times on machine M_B that does not depend on scheduling, and $I(\pi) = I_1 + I_2 + \cdots + I_n$ is the total sum of idle times and is expressed by

$$I(\pi) = \max_{1 \leq j \leq n} \left\{ \sum_{h=1}^{j} p_A(h) - \sum_{h=1}^{j-1} p_B(h) \right\} \qquad (8.3)$$

This formulation looks complicated, but an optimal sequence with the minimum idle time (thus, the minimum makespan) can be obtained by a simple rule, that is, if

$$\min\{p_A(j), p_B(k)\} \leq \min\{p_A(k), p_B(j)\} \tag{8.4}$$

then job j is optimally scheduled earlier than job k. This is realized by a simple procedure: find the shortest processing time among the $2n$ processing times (breaking ties arbitrarily). If it belongs to machine M_A, then put the job on the first place in the sequence, otherwise on the last place. Repeat this procedure for the remaining jobs until all jobs are scheduled.

Note that the makespan can also be expressed by the critical path on which the machines keep busy. That is,

$$F_{\max}(\pi) = \max_{1 \leq j \leq n} \left\{ \sum_{h=1}^{j} p_A(h) + \sum_{h=j}^{n} p_B(h) \right\} \tag{8.5}$$

This expression will be referred to in Section 8.4 in which an unlimited buffer 2-machine FSP with transportation is discussed. Let us consider an example of processing the 5 jobs in Table 8.1 on this two-machine flow shop. In Fig. 8.2, we show two example schedules; here the horizontal axis is the time, and the two machines M_A and M_B are shown vertically (this is called a Gantt chart). Schedule (a) is according to the job numbers, which results in idle time 8 and makespan 20. Schedule (b) is an optimal schedule with the shortest makespan 15.

In 1954, Johnson [38] formulated Equation (8.4), which is considered the birth of the discipline of scheduling research.

We consider another 2-machine FSP model in which no buffer between machines exists. In this case a job is blocked (or interferred), $i.e.$ it cannot leave machine M_A immediately after being finished (unlike the unlimited buffer FSP), if the preceding job is not finished on machine M_B. The schedule of sequence $\pi = (1, 2, \cdots, n)$ is given by

$$
\begin{aligned}
s_A(j) &= s_B(j-1), \\
f_A(j) &= s_B(j-1) + p_A(j) \\
s_B(j) &= \max\{f_A(j), f_B(j-1)\} \\
&= s_B(j-1) + \max\{p_A(j), f_B(j-1) - s_B(j-1)\} \\
f_B(j) &= s_B(j) + p_B(j) \\
&= s_B(j-1) + \max\{p_A(j), p_B(j-1)\} + p_B(j) \\
&= f_B(j-1) + \max\{p_A(j) - p_B(j-1), 0)\} + p_B(j), \\
&\qquad j = 1, 2, \cdots, n
\end{aligned} \tag{8.6}
$$

where $s_B(0) = f_B(0) = p_B(0) = 0$. Job j is blocked on machine M_A, if $f_A(j) < f_B(j-1)$. Then, the recursive equation 8.6 leads to the makespan

$$F_{\max}(\pi) = f_B(n) = \sum_{j=1}^{n} \max\{p_A(j) - p_B(j-1), 0\} + \sum_{j=1}^{n} p_A(j)$$

and to the idle time

$$I(\pi) = \sum_{j=1}^{n} \max\{p_A(j) - p_B(i-1), 0\} \tag{8.7}$$

where $p_B(0) = 0$.

In Fig. 8.3 we show two schedules for the same example as Table 8.1. Schedule (a) is according to the same job sequence as Fig. 8.2b, where job 4, the second job, is blocked on machine M_A and the makespan results in 16. Schedule (b) is an optimal one with makespan 15.

In 1956 Gilmore and Gomory [39] solved a special case of the traveling salesman problem of which the outline is as follows: There are $(n+1)$ cities $0, 1, \cdots, n$ that the salesman travels to, each having two state variables, A and B. The distance from city i to city j is given by

$$d(i,j) = \int_{B_i}^{A_j} f(x)dx, \quad \text{if } B_i \leq A_j$$

$$= \int_{A_i}^{B_j} g(x)dx, \quad \text{if } B_i > A_j \tag{8.8}$$

where $f(x) + g(x) \geq 0$. The task is to find the minimum-length tour.

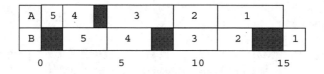

(a) Schedule according to the same
    job sequence as Fig.8.2(b)

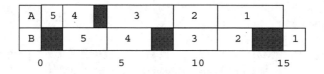

(b) Schedule with the minimum makespan

Fig. 8.3. Schedules for the no buffer 2-machine flow shop

Let

$$f(x) = 1, \quad g(x) = 0, \quad A_j = p_A(j), \text{ and } B_i = p_B(i),$$

then

$$\begin{aligned}
d(i,j) &= \max\{p_A(j) - p_B(i), 0\}, \quad i,j = 1,2,\cdots,n, i \neq j \\
d(0,j) &= p_A(j), \quad j,j = 1,2,\cdots,n \\
d(i,0) &= p_B(i), \quad i,j = 1,2,\cdots,n
\end{aligned} \tag{8.9}$$

Let a tour be $\pi : 0 \to 1 \to \cdots \to n \to 0$, then the total tour length, i.e., $d(0,1)+d(1,2)+\cdots+d(n,0)$ is expressed by $I(\pi)$ of Equation (8.8). Thus, the Gilmore and Gomory algorithm for the traveling salesman problem can solve the no-buffer two-machine FSP. However, the algorithm is too sophisticated to introduce in this limited space, although Equation (8.8) looks simple.

These two-machine flow-shop scheduling problems are extreme with respect to buffer space. In most real factories, the buffer space in between machines is finite; and more than two machines exist. Such more practical situations, however, make problems intractable (so called NP-hard) as shown below.

8.2 Classification of Jobshop Scheduling Problems

Different job-shop scheduling problems arise when we have to consider different constraints on the execution of jobs, or different objective functions to be optimized. Constraints can include a semiordered relationship between jobs (*i.e.* precedence constraint), job release time constraint for jobs, setup time constraint between jobs and others. The objective function can be not only the minimization of the makespan as discussed above, but also for instance the minimization of the tardiness (the overruns of deadlines), the mean flow time (the mean sojourn time of the jobs in the shop) that aims at the minimization of the WIP (work-in-process) and so on; these objectives may be competitive and call for different optimal schedules.

JSP is classified according to the triplet $(\alpha|\beta|\gamma)$, consisting of the structure of the machines (denoted by α), the constraints on the jobs (denoted by β), and the objective function (denoted by γ). Examples of the symbols used for each position are listed below.

Item	Symbol	Meaning
α	1	1SP
	P	PSP
	F	FSP
	O	OSP
	J	JSP
β	prec	Precedence constraints
	r_j	Release time constraints
γ	C_{max}	Total completion time (or makespan)
	L_{max}	Maximum lateness

For instance, $1|r_j|L_{max}$ means that under release-time constraints, we want to minimize the maximum lateness (L_{max}) for a single-machine job-shop. As another example, the above unlimited buffer two-machine flow shop problem is written as $F2||C_{max}$; here $F2$ means two-machine flow-shop and the no-buffer one as $F2|\text{no buffer}|C_{max}$.

8.3 Computational Complexity and Optimization Methods for JSP

The solution set of job-shop scheduling problems (i.e. the set of different feasible schedules) has different size, depending on the problem size, which in turn depends on the number of jobs n and the number of machines m; but in any case, it is finite. Thus in principle if we enumerate the feasible solutions, we will find the optimum; however, for large problem instances this is not possible in practice. For instance, in the case of the most simple 1-machine problem, the dimension of the solution space is on the order of $n!$. In the case of $n = 10$, there are roughly 3.6 million schedules, which may not be an inordinately large number for solving by modern computers. However, for $n = 100$ (which in fact is not even a really large number for practical problems), $n! \approx 10^{159}$, a number that is not tractable even in millions of years on supercomputers. This abrupt increase in computing costs is called *combinatorial explosion*.

Naturally, there are problems for which more efficient procedures (algorithms) are available (*e.g.* the Johnson algorithm and the Gilmore–Gomory one), so they are easily solved. Finding such algorithms is the basic task of scheduling research, which at the same time is also an intellectually satisfying activity (in fact, even without specialist training, it can happen that one finds a new algorithm when thinking about a scheduling problem, quite like when finding the solution to a puzzle). However, there are problems where we are forced to abandon the hope for finding such algorithms. In fact, from the viewpoint of computational complexity theory, while there are problems where the computational cost of the solution grows polynomially with the problem size, there are also problems where the solution cost in the worst case has a combinatorial explosion. These intractable problems are equivalent to each other in the sense that there is a class of combinatorial problems (called the *NP-complete* class for *yes* or *no* decision and *NP-hard* for optimization) in which if any one can be solved in polynomial time, the remaining can also be done efficiently. But there is a huge number of such equivalent problems that remain unsolved so far (see [40] for an excellent guide to NP-completeness).

For instance, the above Johnson problem $(F2||C_{max})$ and the Gilmore-Gomory problem $(F2|no\ buffer|C_{max})$ are solvable in polynomial time of the number of jobs, but $(F3||C_{max})$, $(F2|finite\ buffer|C_{max})$ and $(F2||L_{max})$ are NP-hard. Unfortunately, most of the scheduling problems occurring in practice belong to the latter class (see *e.g.* [41] for the computational complexity of scheduling problems). Therefore, the next task is to decide how to handle such difficult problems.

The approach of formulating the problem as a mathematical model, then attacking it by mathematical analysis, looking for either an exact or an approximate solution, is called the OR (operations research) approach, for historical reasons. One of the canonical OR methods of solving NP-hard problems exactly, is the class of enumerative approaches, exemplified by the *branch-and-bound* method (see *e.g.* [42] and [43] for enumerative approaches to combinatorial optimization). In spite of the above computational complexity analysis, it is a surprising fact that the NP-hard $(F3||C_{max})$ problem with 1000 jobs can be solved in reasonable time with a better than 99% success rate, with modern branch-and-bound methods [44]. This is attributable to the development of such algorithms that can keep the occurrence of the worst cases, predicted by analysis, to a very low rate.

Another approach, based on discovering the particular properties of a problem, and by meticulous design finding the exact or very close approximate solution in a short time, is called *heuristics* from the Greek root of "discovery" (see *e.g.* [45] for an introduction to heuristic scheduling methods). In contrast to this, *metaheuristics*, a methodology recently receiving much interest, discards the strong dependence on the particular characteristics of each problem, and instead tries to find generally useful methods (see *e.g.* [46] for comprehensive explanation of metaheuristics). These methods, represented by *simulated annealing* (SA), *genetic algorithms* (GA), and *tabu search* (TS), require large amounts of computation, which, however, is not necessarily a serious disadvantage nowadays with the quick advances in computing; on the other hand, they will sometimes outperform even heuristic methods.

In the case of practical scheduling problems, we often encounter constraints that are complex and not amenable at all to analytical modeling (ill-structured). In such cases, the knowledge of experts (their know-how), obtained by long experience, often yields better results than mathematical techniques. The approach of collecting such human knowledge, and using it in computational solutions, is the field of *artificial intelligence*(AI). (After Fox [47] applied AI methods to practical scheduling problems in the 1980s, it provided an impetus to the development of this field [48]). The well-known *expert systems*are a representative of AI methods used in scheduling problems. We can say thus that AI methods are used primarily not so much to find the optimum, rather to generate feasible solutions. For example, in the case of production scheduling, when there is a machine breakdown or other nonpredictable event, the original schedule can no longer be executed, and thus it must be revised immediately (reactive scheduling). In such cases AI methods are often effective.

The most practical method, which does not need expert knowledge and can be flexible in complex real-world situations, is simulation. In the case of simulation, the research target is finding models that can flexibly reflect the actual situation, and evaluating scheduling rules. In recent years, production systems have been revolutionized by advances in automation and information technology; thus the know-how obtained with existing production systems might no longer be relevant. In such cases, it is often useful to train first with simulated systems, and then apply the obtained know-how to actual systems.

An optimization technology that does not fit into the above classification is the *theory of constraints* (TOC) developed by Goldratt [49]. This is called the "secret weapon" behind the reconstruction of the American manufacturing industry. Originally it was based on the solution methodology *optimized production technology* (OPT) that was developed for actual production-scheduling problems.

8.4 Robotic Cell Scheduling Problems

Robots are used for automatic material-handling operations. A robot loads items at a location (e.g., a warehouse) and transports them to another location (a production line) in a facility (a factory). A typical example of a robot is an AGV with a robot arm. A robotic cell is a FMC (flexible manufacturing cell) that is a small-scale FMS (flexible manufacturing system) in which one or a few robots serve between close locations. Robotic cells for manufacturing including assembling have been used worldwide in many industries, especially automotives and electronics, to make manufacturing more flexible and efficient for multiproduct small-sized production since the early 1980s (*e.g.* see [50]). For example, a Japanese factory [51] increased its cell production from 90 motors a day when manually served, to 300 motors by introducing robots in 1982. Scheduling problems for such robotic cells, generally become more complicated than the corresponding job shop scheduling problems (see [52] and [53]). This section discusses a robotic scheduling problem that correspond to the two-machine FSPs with buffer.

A robotic cell of flow shop type, as shown in Fig. 8.4, consists of two machines M_A, M_B on which jobs are processed sequentially, a robot R and an intermediate station IS. Robot R unloads a job from M_A, sends and loads it to IS, sequentially, and then deadheads. Station IS can process an unlimited number of jobs simultaneously (*e.g.* cooling at an open space). This is an extension of the above-mentioned two-machine flow shop with an unlimited buffer where no robot is considered. A typical behavior of the robotic cell with 3 jobs is illustrated by the Gantt chart in Fig. 8.4. Job 1 is served without interruption. Job 2 waits (or it is blocked) on M_A until R returns. Job 3 waits at IS until job 2 is finished on M_B. Two jobs 1 and 2 are concurrently served at IS.

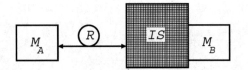

Fig. 8.4. A robotic cell with an intermediate station

The following parameters are considered for scheduling:

$u_A(j)$ $(j = 1, 2, \cdots, n)$: unloading time of job j from M_A.
$p_i(j)$ $(i = A, B, \quad j = 1, 2, \cdots, n)$: processing time of job j on M_i.
l_{IS}: constant loading time into IS.
$c_{IS}(j)$ $(j = 1, 2, \cdots, n)$: cooling time of job j at IS
t_R: constant transportation time from M_A to IS.
d_R: constant deadhead time of R from IS to M_A.

Let $\pi = (1, 2, \cdots, n)$ be a sequence of releasing n jobs and

$$P'_A(j) = p_A(j) + u_A(j) + c_{IS}(j), \quad j = 1, 2, \cdots, n$$
$$P_A(j) = P'_A(j) + \max\{0, t_R + l_{IS} + d_R - P_A(j)\} \quad (j = 2, 3, \cdots, n)$$
$$C_{IS}(j) = t_R + c_{IS}(j)$$

then the start (of processing) and the finish (of unloading) times of job j on M_A are, respectively, given by

$$s_A(j) = f_A(j - 1),$$
$$f_A(j) = s_A(j) + \max\{p_A(j), t_R + l_{IS} + d_R\} + u_A(j)$$
$$= P'_A(1) + \sum_{k=2}^{j} P_A(h), \quad j = 1, 2, \cdots, n \tag{8.10}$$

where $f_A(0) = 0$. Arrival and departure times of job j at IS are, respectively, given by

$$s_{IS}(j) = f_A(j) + t_R$$
$$f_{IS}(j) = s_{IS}(j) + c_{IS}(j), \quad j = 1, 2, \cdots, n \tag{8.11}$$

Start (of processing) and finish times of job j on M_B, are, respectively, given by

$$s_B(j) = \max\{f_{IS}(j), f_B(j-1)\}$$
$$f_B(j) = s_B(j) + p_B(j), \quad j = 1, 2, \cdots, n \tag{8.12}$$

Let

$$P_B(j) = p_B(j) + C_{IS}(j), \quad j = 1, 2, \cdots, n$$

then, the makespan results in

$$F_{\max}(\pi) = P_A(1) + \max_{2 \le j \le n}\left\{\sum_{h=1}^{j} P_A(h) + \sum_{h=j}^{n} P_B(h)\right\} - \sum_{j=1}^{n} C_{IS}(j) \tag{8.13}$$

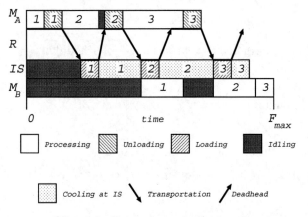

Fig. 8.5. Gantt chart of robotic cell

Here, refer to Equation (8.5), the makespan in the case of no transportation. Then, it is obvious that this robotic scheduling problem can be solved by repeating Johnson algorithm n times, changing the first job.

Further extensions of the robotic cell problem include the case of no buffer, and other special cases. The reader is encouraged to consider formulating transportation problems in the framework of the scheduling model, which might make possible the application of an existing solution.

Part III

Intelligent Control Methods for
Transportation Systems

9

Analytical and Heuristic Control of Transportation Systems

This chapter will introduce some actual control methods, starting from a historical perspective. The topics covered reflect the authors' personal experience, and therefore here we shall deal almost exclusively with elevator group control systems. However, the methods themselves are not restricted to elevators, and in fact have been used in many other fields.

9.1 Evolution of Control Methods

The goal of elevator group control can be stated in general terms as a multi-objective optimal control problem:

When given

- an elevator system with known system parameters,
- a set of observable state values,
- a set of decision parameters for control,
- and a set of control objectives,

find the *policy* (a mapping from states to decisions) that can calculate at each decision instant a decision vector, such that the control objectives are optimized in a sense (*e.g.* Pareto).

Usually, the control objective is understood as a combination of the following functions:

- passenger service level (distribution of waiting times and total service times),
- total traffic throughput (handling capacity),
- energy consumption,
- psychological objectives (immediate announcement of the allocated car, avoiding overcrowding in cars, *etc.*),
- special objectives (*e.g.* avoiding the noise caused by several cars running in parallel in connected elevator shafts, *etc.*).

However, the most important objective by far is the passenger service level, which is closely related to the average waiting time (AWT), and the probability of long waiting (LWP). A further important and very visible parameter is the percentage of announcement errors of allocations. This is the probability that a hall call, which had been allocated to a given car, is served by another car. It becomes important in the case of group control systems with an *immediate announcement* feature, where the initial allocation is announced to waiting passengers; they will be frustrated when that announcement proves to be false. In this work, we will consider only these values for evaluating the performance of elevator control algorithms.

Historically, elevators had been used in groups from early times in larger buildings, making it necessary to consider the problem of group control. Although for a long time elevators were operated manually, and the operators used their experience for dividing the work among the cars, human operators were eventually replaced by automatic group control operation. Elevator group control algorithms have evolved in several stages during the last 30 years. The main phases are summarized below, with brief discussion of their characteristics.

9.2 Analytical Approaches

From the point of view of control engineering, we would call the problem of controlling transportation systems "solved", if we could find an appropriate analytical model for our system, with an explicit, closed-form solution.

Unfortunately, most of the systems in use are far too complex for the analytical approach to succeed. Still, some work has been done in this direction.

One of the earliest algorithmic approaches is found in Knuth's classic work [54], who noticed that Karp's "one-tape sorting algorithm" can in fact be considered as the optimal solution of a stylized elevator route planning problem.

At about the same time, Levy and coworkers [55, 56] created a drastically simplified model of the elevator group, with a relatively small discrete state space. The behavior of this system could then be modeled as a Markov chain, and solved semianalytically by linear programming.

A further, still-continuing branch of the analytical approach is the *competitive analysis* represented by the work of Krumke [57] and others [58]. In this research, the performance of specific algorithms for transportation problems, that are *realizable*, *i.e.* use only information available for a controller, is analyzed in comparison with such idealized algorithms that have full knowledge of the future. The term *competititve* refers to the property of a realizable algorithm, that its performance is poorer at most by a constant factor. The attractive feature of this approach is the strict analytical bounds that are obtained for the performace. To our knowledge, there are still no such algorithms that are both amenable to this analysis, and at the same time offer acceptable performance in practice. However, there is steady progress, and the reader might want to get acquainted with this fascinating research area.

9.3 Heuristic Rules

Much of the history of the elevator control systems had been spent on finding "good" control rules that could be implemented by the existing control equipment. Until about 1970, this had meant a few stepping counters and a few dozen relays, which was all that could realistically be dedicated to automate the elevator control. Thus the rules that we review here had to be designed with these limitations.

In spite of the low capabilities of the hardware, the inguenity of the control engineers has still succeeded in finding some basic tricks that did work to a surprising extent; for a quite thorough exposition see Barney [5]. We list here only a few:

- **Multiple selective-collective service**
  A traditional approach is to use a "greedy" strategy. The most simple way of serving the same floors with several cars is by using the heuristic rule:

  The nearest car takes the call.

  Thus each car keeps running in a direction, stopping to pick up any waiting passengers, and delivering them to their destination floor. This is the so-called "selective-collective group operation" that is still used sometimes. This seemingly obvious algorithm, however, has poor performance. Hence, it was soon abandoned and the search for effective group control started and continues today. One main reason for the failure of this algorithm is the phenomenon of *bunching* [5], *i.e.* the tendency of cars to synchronize the phase of their movement, and travel roughly together. This means that the average waiting time becomes proportional not to $INTV = \frac{RTT}{m}$, but to RTT.

- **Terminal dispatching**
  As a variation on the above heuristics, we can try to keep the cars waiting at the lower terminal floor, until a predetermined dispatching interval has passed, in order to equalize their workload, and keep them from "bunching". Terminal dispatching means the starting of cars in predefined equal intervals at the lower terminal floor. This algorithm usually operated during the *up-peak* period. When a car arrived at the terminal floor, passengers could board, but the car waited with open doors until the *dispatching signal*. Dispatch intervals were set so that cars would leave nearly full, in approximately equal intervals. Low hardware requirements were probably one of the main reasons for selecting such a simple algorithm at the time of its introduction. Also, from Equation (6.2) it is clear that for a given mean service interval, the expected waiting time is minimized if the variance of

the service time is minimal. The terminal dispatching method was motivated by this consideration; however, it works well only for the *up-peak* traffic. As one of its drawbacks, the idling at the terminal is inefficient and irritating to passengers. Even the closely related *multiprogram dispatching* algorithms, which used some special heuristics for other traffic patterns, could not overcome the basic limitations of the dispatching method.

- **Zoning control**
  Zoning control means the division of the building into zones, and assigning cars to serve a given zone. This method is still used with variations [5]. Its main heuristics are:
  - divide the building vertically into contiguous zones, and assign cars to serve calls in each zone;
  - to keep cars apart (prevent *bunching*) we limit the number of cars in each zone, similarly to railway systems;
  - prioritize zones by the elapsed waiting time, thus try to limit the maximum waiting time;
  - zoning also implies an attempt at reducing the travel distance and number of stops of each car, by serving closely spaced calls by the same car

  However, determining the zones by heuristic rules is rather arbitrary, and a natural next step was considering each floor (each call) as a separate *zone*.

There are many variations on these themes, and we are forced once again to direct the reader to Barney for a fascinating exposition of these techniques.

9.3.1 Algorithmic Control

As soon as minicomputers and later microcomputers appeared, elevator engineers eagerly started putting them to work as group controllers. This started a proliferation of newer and newer heuristic methods; however, in spite of the considerable effort, the progress was still rather slow. One main reason is that even for an expert, it was found to be very difficult to describe in strict logical rules the knowledge that was gained from experience.

The following approaches are representative of this stage.

- **Call allocation**
  By call allocation, we mean the allocation of each new hall call to a car, based on an estimated merit function. In this method, instead of zones, each individual hall call had a car allocated, with the allocations being determined by an *evaluation function*. The reasoning behind this algorithm is simple: If we could predict the future, the elevator group control problem would be conceptually very easy. We would define a penalty function $F(\{\mathbf{c}(t)\}, \{a(t)\})$ to be minimized, where $\{\mathbf{c}(t)\}$ is the traffic, and $\{a(t)\}$ is the sequence of decisions (allocations). Then for each possible sequence of car allocations, we could calculate F, and select

$\{\hat{a}(t)\} = \arg\min_a F(\{\mathbf{c}(t), a(t)\})$. However, in reality we need to decide each $a(t_1)$, when the later traffic $\mathbf{c}(t_2 > t_1)$ is not yet known. The simple solution taken in the *call-allocation method* is to use a heuristic estimate of F for each possible allocation, and choose the one with the lowest cost. The main problem with this method is that the quality of this estimate determines the efficiency of the control.

The call-allocation method made it possible in principle to allocate a car immediately to a new call, and announce the allocation to the waiting passengers. This allows them to prepare for boarding, helps the orderly formation of queues, and reduces congestion by keeping *up* and *down* queues apart. However, the allocations often had to be changed, since the evolving elevator situation made it necessary to revise the decisions. Because of this, immediate announcements were not widely used with these group controllers. The difficulty of improving the evaluation function, and the poor performance of the *immediate announcement* feature led to the development of the AI group control algorithms.

- **The ACA (adaptive hall call allocation) algorithm**
  Elevator equipment behaves essentially in a deterministic way. Randomness comes from the unpredictable behavior of the users. However, if at a given instant the destination of each passenger is known, and forgetting for a moment that others will continue to come, we can express explicitly the control objective, and try to select a decision that optimizes it. For instance, in the notation of Barney, let us express AWT (average waiting time) for a set of $N(I)$ existing calls for each of L cars, and one new call allocated to car K, as

$$AWT = \frac{NWT(K) + \sum_{I=1, I \neq K}^{L} WT(I)}{1 + \sum_{I=1}^{L} N(I)}$$

As a variant of the "call allocation" method, we can then select the allocation K^* that minimizes AWT. Similar expressions can be written for other objectives, like AJT (average journey time), *etc.*

These methods were successful, but development did not stop here.

9.3.2 Fuzzy AI Group Control

The main motivation for further development came from the engineers, who wanted some way of incorporating their expert-domain knowledge into the elevator-group controllers.

This knowledge consisted of two main kinds of rules. The strategic rules were insights into how to handle typical traffic situations; for instance, what to do to prevent unequal service or bunching. The tactical knowledge related to the details of control, like how to handle special calls for handicapped persons, *etc.*

A very successful solution to the problem of encoding expert knowledge in a computational form had been proposed by Zadeh, with the introduction of "fuzzy logic" , that made possible the implementation of "linguistic rules".

The theory and application of fuzzy logic is described in many readily available textbooks and monographs, so we will not try to give a complete exposition, only indicate the general way of thinking in broad terms.

By fuzzy AI group control, we mean a version of the call-allocation method, which uses expert rules in the form of a fuzzy knowledge base [59, 60]. "Experts" prepared control rules for the elevator group controller, by considering various situations in simulation experiments and during actual operation in buildings [61]. Some of these rules apply to particular traffic situations, like e.g. "*In down peak, try to run some cars to lower floors directly*", others are general rules, like e.g. "*If two cars are near, try to have the tailing car take more calls, in order to separate the cars*". The antecedents (inputs) of the fuzzy rules are built from fuzzy variables, *i.e.* fuzzy sets. These reflect the state of the elevator system, in so-called linguistic form; *e.g.* two cars are very near, or the traffic has Down Peak pattern, *etc.* From the fuzzy input variables, the fuzzy rules output fuzzy truth values. The fuzzy inference system combines these truth values, and generates fuzzy decision values, which are finally *de-fuzzified* (usually by weighted means). In the case of the FLEX-8800 system of Fujitec [59, 60], the inference system is further enhanced by the use of multiple rule bases, which are applied sequentially, until a conclusive decision can be made.

Fuzzy group-control systems achieved high performance, both by reducing the passenger waiting times, and by reducing the prediction error rate for immediate announcement of car allocation. However, the fuzzy systems had no capability of adapting their rules to the specific situations of each building, so if the generic control does not fit, they must be tuned manually. It also means that they cannot adapt automatically, if the conditions change in a building. These shortcomings were addressed by the development of the neurocontrolled group control systems.

9.4 Early Approaches to Optimal Control

In this section we briefly review some of the earlier approaches to the development of optimal control methods for transportation problems. We include this section here, since these works were contemporary with the heuristic approaches. More advanced optimal control methods will be covered in the following chapters.

The thesis of Closs [11] deals with the optimization of the operation of the elevator system. This includes the response of a car to the calls, and also the allocation of calls to cars. Because of the limited computational resources at the time of that study, only small-scale tests were possible. Although Closs has proposed solving the optimal allocation problem for already-registered calls, it seems that instead of optimization in the strict sense, mostly simulation studies were done. He reports, however, finding that the S/C algorithm is nearly optimal for a single car. Closs also recommends using destination-floor registration, which, its merits notwithstanding, is, however, still quite uncommon even today.

Levy and coworkers [55, 56] have also considered the optimal elevator group control problem, using a dynamic programming approach. They represent the state of a simplified model of the elevator system by a relatively low-dimensional discrete-valued vector $\mathbf{b}(t)$. The operating policy g is defined in the form of a mapping of system states to control decisions:

$$g \; : \; \mathbf{b} \mapsto \mathbf{d} \tag{9.1}$$

where $\mathbf{d}$ determines the state-transition probabilities in the next time step. The cost function to be minimized is defined as the length of the *busy period* $K(\mathbf{b}(t), g)$, *i.e.* the expected interval from t until all queues in the system become empty for the first time. The optimal policy g^* corresponding to the steady state solution is found by solving Bellman's equation

$$K(\mathbf{b}, g^*) = \min_{g \in G} \left(\chi_\beta(\mathbf{b}, g) + \int_{(\beta)} K(\beta, g) dF(\beta | \mathbf{b}, g) \right) \tag{9.2}$$

where $\chi_\beta(\mathbf{b}, g)$ is the transition time from the state $\mathbf{b}$ to the state β, and F is the probability distribution of β, conditioned on the initial state and the control policy. In Levy's method, Equation (9.2) is solved numerically, by using a discrete state vector $\mathbf{b}$, with binary-valued components indicating the relative position of the elevator cars and the nearest call, the presence or absence of other calls, *etc.*; by using binary vectors, it becomes possible to further reduce sets of states by Boolean simplification methods.

A drawback of this method is the assumption of a memoryless controller with discrete, low-dimensional state space. This requires a very drastic simplification of the state of the elevator system, discarding continuous variables like elapsed waiting times, *etc.* A further problem is the requirement of complete *apriori* knowledge of the traffic densities and the O/D distributions. This makes it difficult to apply the method to an actual online control, which would require adaptation to changing traffic conditions. In practice, a group controller would have to estimate the traffic parameters from observed data, and modify the control parameters accordingly, thus a precalculated policy is not appropriate.

This method also has severe restrictions on the allowable policies, most notably the inability to determine the allocated car at the time of hall-call registration. In Japan, modern elevator group controllers announce immediately the allocation of hall calls, to inform the passengers. This requires fixing of the allocations, unlike in the case of Levy's method.

More recently, advances in combinatorial optimization by the *simulated annealing* [62], *threshold accepting* [63], or *genetic algorithm* [64] methods have made it possible to use these methods for the group control problem [65, 66].

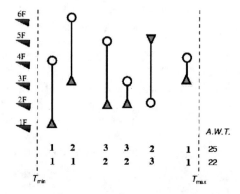

Fig. 9.1. A deterministic combinatorial optimization approach to the elevator group-control problem. In a given interval, all calls are assumed known, and all control decisions are made simultaneously.

As a simplified case, it is possible to consider a noncausal model, *i.e.* a case where complete information information is available in advance for all decisions. We assume that all decisions are made simultaneously, and the objective function is minimized by explicit search in the decision space $D^n = \{1, ..., m\}^n$, as

$$\min_{C \in D^n} H(C, \mathbf{U}(t_n)) \qquad (9.3)$$

Here, $\mathbf{U}$ is the set of system state vectors between time $T_{\min}$ and $T_{\max}$, up to the nth passenger (or call) at t_n. C is the set of decisions corresponding to each passenger, *i.e.* the numbers of the car allocated to that passenger. See Fig. 9.1 for an explanation, where two possible decision sets, together the corresponding H (average waiting time) is shown. Obviously, an actual controller must operate causally (without knowledge of future inputs), but even the simple noncausal model can be used, *e.g.* to obtain absolute performance limits for a given traffic situation.

There have been various proposals for solving the combinatorial optimization problem of Equation (9.3), using SA (simulated annealing) [67], or GA (genetic algorithm) [66] methods. An improved solution, using the TA (threshold accepting) method [65], has been proposed by one of the authors. The TA method has been proposed [63] as an efficient alternative to the SA method, based on a modification of the acceptance probability during the search in the solution space. The method of [65] takes into account one characteristic of the elevator system, *i.e.* that the optimal service will probably distribute the calls (or passengers) equally between the cars.

We call a solution (a set of decisions) *balanced*, if the difference between the number of passengers (or calls) allocated to any two cars is minimal. In this method, we start the search from a balanced allocation state, and maintain the balance through the search, by swapping allocations between cars (see

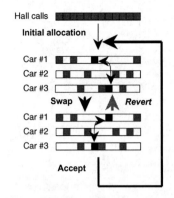

Fig. 9.2. Search in the restricted (balanced allocation) subspace. Interchange of two calls between two cars keeps the allocations balanced.

Fig. 9.2). This results in the reduction of the search space, and the smoothing of the *energy landscape*, thus allowing the use of the efficient TA algorithm, which presupposes that the energy function along the search trajectory has no large jumps. The TA method can only escape from local minima if at all intermediate states the worsening of the goal function is below a preset margin. We can expect that this condition will be satisfied, due to the maintenance of the load balance among the elevator cars [1].

Some extensions have also been proposed to overcome the shortcomings of the deterministic model; for instance, by using a set of simulated random continuations at each decision time, and optimizing over the union of known (past) and simulated (future) decisions [68] (see Fig. 9.3). We call the union of the given allocation problem and a simulated allocation problem a *random continuation*; for instance, in Fig. 9.3 the first 3 calls are *real*, but they are extended with three versions of randomly simulated, artificial allocation problems. We solve all the *random continuations*, and select those decisions for the *real* problem, which were successful in the majority of cases. This approach could be used in principle for online control: at each decision instance, the existing calls would be extended in a few simulated versions into the future, and after solving these *random continuations*, the most frequent decision for the present call is selected as the result. However, at present the computational requirements of the combinatorial optimization phase are too high to use this approach.

---

[1] The reason is that although individual calls will be served faster or slower, the overall load of each car remains invariant, so the overall service level also changes only by small amounts. If the load balance were allowed to change, gradually shifting calls to a particular car would make that car go around slower, thus resulting in a fast change in the overall service level.

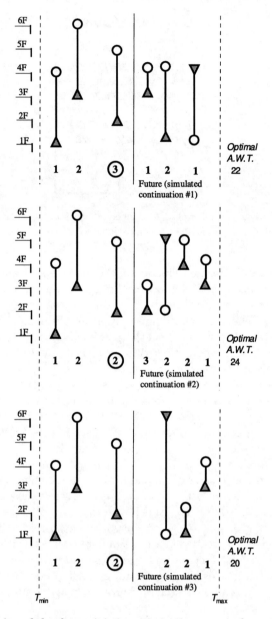

Fig. 9.3. Extension of the deterministic optimization approach to random continuations of the state of the system

10

Adaptive Control by Neural Networks and Reinforcement Learning

10.1 Information Processing by Neural Networks

Information processing by *neural networks* (NN) is based on modeling the information processing of the neurons of animal brains. They have the following structure:

- there are nonlinear processing units, modeling the biological neurons, which are interconnected,
- some of them receive external inputs, while the outputs from some others constitute the system output,
- the interconnections correspond to the synaptic connections between neurons, and its strength is made variable; this becomes the adjustable parameter of the NN,
- by the adjustment of this parameter according to examples, which is called learning, the NN obtains the required information processing functionality.

The main characteristics of information processing by NNs are:

- analog processing;
- parallel processing;
- function acquisition by learning.

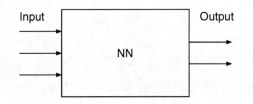

Fig. 10.1. Information processing by neural networks

There had been many proposals for NN architectures and learning methods, but the two most representative examples are the multilayer perceptron (MLP), and the self-organizing map (SOM).

10.2 Multilayer Perceptrons

10.2.1 Model of the Processing Units

Let us consider N processing units with x_i outputs. In the processing units of the MLP, the weighted sum of the outputs of other processing units is processed as

$$s_i = \sum_{j=1}^{N} w_{ij}x_j + w_{i0} \tag{10.1}$$

$$x_i = f(s_i) \tag{10.2}$$

Here, s_i is the intermediate output signal of the unit (corresponding to the membrane potential of biological neurons). For the nonlinear function $f(x)$, the

$$\text{sigmoidal function} \qquad f(x) = \frac{1}{1 + \exp(-x)} \tag{10.3}$$

is often used. This can be interpreted as the approximation of the

$$\text{step function} \qquad f(x) = \frac{\text{sgn}(x) + 1}{2} = \begin{cases} 0 \text{ for } & x < 0 \\ 1 \text{ for } & x \geq 0 \end{cases} \tag{10.4}$$

by a differentiable function. This makes possible the *backpropagation algorithm*, which will be introduced later. The linear function

$$f(x) = x \tag{10.5}$$

is also used. If we introduce an artificial unit x_0 with an identically 1 output, we can write Equation (10.1) formally without a constant term:

$$s_i = \sum_{j=0}^{N} w_{ij}x_j \tag{10.6}$$

10.2.2 Structure and Operation of the Multilayer Perceptron

The *multilayer perceptron* (MLP), as mentioned in Section 10.2.1, has its information flow in a single direction, from the input units to the output units.

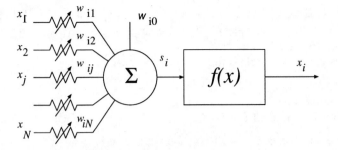

Fig. 10.2. Processing Unit

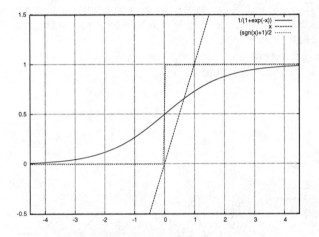

Fig. 10.3. Characteristics of the Unit

- Units that just receive an input signal and forward it as their output, are called *input units*
- units that provide the system output signal are called *output units*,
- and those units that are not directly connected either with inputs or outputs to the external world are called *hidden units*.

Units are organized into *layers*, and we talk about the input layer, output layer, and hidden layers.

If all units have linear characteristics, the processing becomes nothing but a repeated linear transformation, with limited capabilities. If we number the units of the MLP starting from the inputs, and proceed toward the outputs, there will be no connections from higher-numbered units to lower ones; that is, we can say that

$$w_{ij} = 0 \quad \text{for} \quad j \geq i \tag{10.7}$$

If we consider the connections w_{ij} as a matrix, then $W = \{w_{ij}\}$ is lower triangular. If, furthermore, we allow connections only between adjacent layers, it also becomes a block matrix.

Let us introduce indices for

- input units as $1, \cdots , I$,
- hidden units as $I + 1, \cdots , I + H$,
- and output units as $I + H + 1, \cdots , I + H + O(= N)$;

then the operation of the MLP can be written as

1. set the outputs of the input units to $x_i, i = 1, \cdots , I$;
2. calculate the value of the hidden and output units in sequence, according to Equations (10.1)–(10.4). Note: units with higher indices have no effect due to $w_{ij} = 0$ for $j \geq i$, which guarantees the feasibility of the calculation.
3. obtain the output of the network as the output values $x_i, i = I + H + 1, \cdots , N$ of the output units.

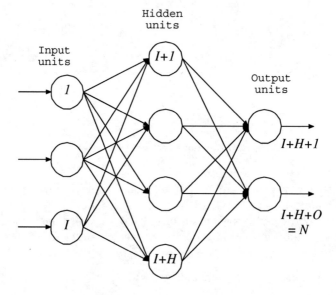

Fig. 10.4. Structure of the MLP

10.2.3 Expressive Power of the MLP

It is known that a MLP with

- (at least) one hidden layer, *i.e.* with 3 layers,

- sigmoidal units in the hidden layer,
- linear units (with unlimited outputs) in the output layer,
- a sufficient number of units in the hidden layer,

is capable of realizing an arbitrary function with any desired accuracy of approximation.

However, we must note that this is a property for which we do have an existence theorem, but it says nothing about whether a set of connection weights can actually be learned to realize a given function.

As a related result, we also know that with binary-valued $\{0, 1\}$ input and output units, with at least 3 layers, with step function units in the hidden and output layers, and with a sufficient number of hidden units, any given logical function can also be realized by the MLP.

10.3 Learning as an Optimization Problem

In the case of MLP, by *learning* we mean the following:

preparing a set of input data and corresponding output data (*teacher signal*), and then adjusting the connection weights w_{ij}, until upon presenting one of the input data to the MLP, the output becomes as close as possible to the teacher signal.

This kind of learning employs a teacher signal, so it is called supervised learning.

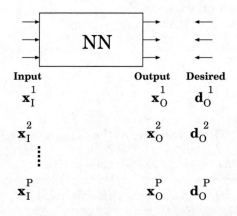

Fig. 10.5. Learning for MLPs

Let us have P pairs of input and teacher signals $(\boldsymbol{x}_I^p, \boldsymbol{d}_O^p)$, $p = 1, \ldots, P$, as shown in Figure 10.5. The many indices are somewhat confusing; we use superscripts for the indices referring to the training data.

Here the input signal $\boldsymbol{x}_I^p$ is a vector with dimension equal to the number of the input units I, and the dimension of the teacher signal is equal to the number of output units O. The factor $1/2$ is introduced to cancel out the factor 2 that we get from differentiating quadratic functions.

For the pth data, we evaluate the error between the output of the MLP $\boldsymbol{x}_O^p$ and the teacher signal, by the squared difference

$$G_p = \frac{1}{2} \sum_{i \in U_O} (d_i^p - x_i^p)^2 \tag{10.8}$$

Here U_O is the set of indices of the output units, and d_i^p, x_i^p are the ith components of the teacher signal and the output values, respectively. Evaluating the above for all P training data,

$$G = \sum_{p=1}^{P} G_p = \frac{1}{2} \sum_{p=1}^{P} \sum_{i \in U_O} (d_i^p - x_i^p)^2 \tag{10.9}$$

gives the error of the MLP; *learning* means minimizing this with respect to the connection weights $\boldsymbol{w}$.

This approach is similar to fitting a function to the observed data by the least squares method; however, usually the fitted function, *e.g.* a linear function $y = w_1 x + w_0$ or a quadratic function $y = w_2 x^2 + w_1 x + w_0$, are linear functions of the fitting parameters w_i, in which case the minimum least square solution is analytically computable. In contrast, the processing units of the MLP have nonlinear (sigmoidal) characteristics, and we have to use numerical methods for the solution.

Thus we can formalize the learning problem as solving the nonlinear optimization problem

$$\min_{\{w_{ij}\}} G = \min_{\{w_{ij}\}} \sum_{p=1}^{P} G_p = \min_{\{w_{ij}\}} \frac{1}{2} \sum_{p=1}^{P} \sum_{i \in U_O} (d_i^p - x_i^p)^2 \tag{10.10}$$

10.3.1 Nonlinear Optimization by the Gradient Method

The basic algorithm of minimizing the function $G(\boldsymbol{w})$ that is nonlinear in the parameter $\boldsymbol{w}$, with respect to $\boldsymbol{w}$, is obtained by giving some initial values to $\boldsymbol{w}$, and then adjusting $\boldsymbol{w}$ in a direction that decreases G.

When G is differentiable, G decreases fastest in the direction opposite to the gradient $-\nabla_{\boldsymbol{w}} G$, which leads to the update rule

$$\boldsymbol{w}^{\text{NEW}} = \boldsymbol{w}^{\text{OLD}} - \alpha \nabla_{\boldsymbol{w}} G \tag{10.11}$$

for $\boldsymbol{w}$, called the *gradient method* or *steepest descent method*. Here, α is a positive value that controls the step size of the update.

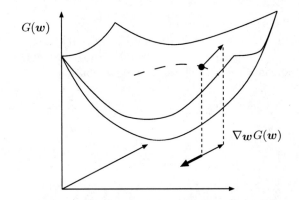

Fig. 10.6. The principle of the gradient method

In the NN, the $\boldsymbol{w}$ vector has two dimensions, but here we reorder it as a single-dimensional vector for readability. Also, we use a column vector for the gradient.

The gradient $\nabla_{\boldsymbol{w}}G$ is the vector obtained by derivating the scalar function G with respect to each component of the vector $\boldsymbol{w}$:

$$\nabla_{\boldsymbol{w}}G = \begin{pmatrix} \frac{\partial G}{\partial w_1} \\ \frac{\partial G}{\partial w_2} \\ \vdots \\ \frac{\partial G}{\partial w_N} \end{pmatrix} \tag{10.12}$$

10.3.2 Derivation of the Learning Rule

Here we will obtain the gradient learning algorithm for a general MLP.

The learning method consists of adjusting $\boldsymbol{w}$ in the direction $-\nabla_{\boldsymbol{w}}G$, since $G = \sum_{p=1}^{P} G_p$, $-\nabla_{\boldsymbol{w}}G$ can be obtained from $-\nabla_{\boldsymbol{w}}G = -\nabla_{\boldsymbol{w}} \sum_p G_p = \sum_p (-\nabla_{\boldsymbol{w}} G_p)$ by evaluating $-\nabla_{\boldsymbol{w}} G_p$ and summing over. This is calculated as follows.

First, for the connection weights of the output units we can simply define

$$\delta_i^p \equiv -\frac{\partial G_p}{\partial s_i} = (d_i^p - x_i^p) f_i' \tag{10.13}$$

[1] and since unit i works as $x_i = f(s_i), s_i = \sum_j w_{ij} x_j$, we obtain

$$-\frac{\partial G_p}{\partial w_{ij}} = -\frac{\partial G_p}{\partial s_i} \frac{\partial s_i}{\partial w_{ij}} = \delta_i^p x_j^p, \quad i \in U_O \tag{10.14}$$

---

[1] δ is the Greek letter *delta*. We use this since we need d for a different purpose.

Here, δ_i^p is called the **error signal**.

For the connection weights to the hidden units (the set of their indices is defined as U_{H}), as the signal is propagated through the MLP to the output units, from the chain rule of differentiation we obtain

$$-\frac{\partial G_p}{\partial w_{ij}} = -\left(\sum_{k \in U_{\mathrm{O}}} \frac{\partial G_p}{\partial s_k}\frac{\partial s_k}{\partial x_i}\right)\frac{\partial x_i}{\partial w_{ij}} = \left(\sum_{k \in U_{\mathrm{O}}} \delta_k^p w_{ki}\right)f_i' x_j^p = \delta_i^p x_j^p, \quad i \in U_{\mathrm{H}}$$

(10.15)

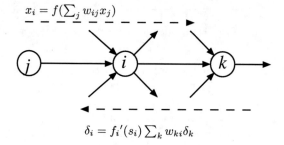

$$x_i = f(\textstyle\sum_j w_{ij} x_j)$$

$$\delta_i = f_i'(s_i) \textstyle\sum_k w_{ki}\delta_k$$

Fig. 10.7. Backpropagation of the learning error signal

Here, the error signal δ_i^p is defined as

$$\delta_i^p \equiv \left(\sum_{k \in U_{\mathrm{O}}} \delta_k^p w_{ki}\right)f_i',$$

(10.16)

and Equations (10.14) and (10.15) can be written formally in the same way. For networks with more layers, this procedure can be applied more times, working toward the input layer. In this calculation, the error signal δ_i^p is propagated from the output of the MLP in the reverse direction toward the input side; from this, the method is called the *error backpropagation* (BP) method[2].

As the most simple variant of the gradient rule, α is set to a positive constant, the so-called *learning coefficient*, and w_{ij} is modified by adding to it $-\partial G/\partial w_{ij}$ multiplied by α, with this performed in a *batch* for all training data:

$$w_{ij}^{\mathrm{NEW}} = w_{ij}^{\mathrm{OLD}} - \alpha\frac{\partial G}{\partial w_{ij}} = w_{ij}^{\mathrm{OLD}} + \alpha\sum_{p=1}^{P} \delta_i^p x_j^p$$

(10.17)

---

[2] By *BP method* we mean both the gradient method, and the learning rules Equations (10.17), (10.18), (10.20), *etc.*

Also, from $G = \sum_{p=1}^{P} G_p$, we can obtain an approximate *online* version of the learning rule by presenting the training data one by one[3]:

$$w_{ij}^{\text{NEW}} = w_{ij}^{\text{OLD}} - \alpha \frac{\partial G_p}{\partial w_{ij}} = w_{ij}^{\text{OLD}} + \alpha \delta_i^p x_j^p \tag{10.18}$$

In the case of a simple implementation with a fixed α, the value of α needs to be determined by experiments. If α is small, the learning will be stable but slow. Conversely, if α is too large, w might oscillate, and the learning will not converge.

Even if the value of α is adjusted, such learning rules are rather slow. As a practical way to accelerate the learning, the **moment method** has been proposed, which reuses the previous correction to the connection weights in the next step:

$$w_{ij}^{\text{NEW}} = w_{ij}^{\text{OLD}} + \Delta w_{ij}^{\text{NEW}} \tag{10.19}$$

$$\Delta w_{ij}^{\text{NEW}} = -\alpha_1 \frac{\partial G_p}{\partial w_{ij}} + \alpha_2 \Delta w_{ij}^{\text{OLD}} \tag{10.20}$$

Here, α_1, α_2 are positive constants, for example with the recommended values of $\alpha_1 = 0.1$, $\alpha_2 = 0.9$.

10.3.3 Hints for the Implementation and Use of the BP Method

There are some points to be kept in mind when implementing and using MLPs. We review them below.

Using MLPs

- MLPs are suitable for wide practical use, not only for pattern recognition tasks, but also for casual data processing, where they are often more efficient than manual data processing by trial and error.
- Due to the characteristics of the component units of MLPs, they are good at finding hidden structures in data. However, the effect of the learned data points can be global, while in some applications we want to restrict the effect of each training data to their vicinity. For such cases, instead of using MLP, it might be better to use k-NN methods [69] or RBF [70] that work according to the distance between the input point and the training data points.
- It is recommended that before trying MLPs with hidden layers, experiments should be made with a network without hidden layers, which is easily trained.

---

[3] Because the connection weights w are modified at each step, the gradient changes, and the rule is only approximate; however, if the learning coefficient is small, this is not a serious issue.

- If pattern recognition is attempted directly on images or speech signals, it puts too high loads on the MLP, and a huge amount of data becomes necessary for training. It is better to employ some proper method for feature extraction, as a preprocessing before the MLP. It should be noted that if the preprocessing of image data includes a thresholding step, the resulting loss of information can be disadvantageous for the operation of the MLP.
- When collecting the training data, it is recommended to cover as much as possible of the existing variations (including things like distortions, impurities, *etc.*).

Structure of the MLP

- In pattern-recognition tasks that require a $0 - 1$ output, sigmoidal units are used in the output layer. For tasks like function approximation, where the range of the output is not limited, the output units are linear $f(s) = s$.
- If the sigmoid function is implemented as Equation (10.3), the internal numerical computation of the exponential function will sometimes overflow. We can multiply the numerator and denominator of the sigmoidal function by $\exp(s)$, and write it as the equivalent $\exp(s)/(1 + \exp(s))$. Depending on the sign of the input, we can switch between the above definitions, thus avoiding the overflow problems.
- As one of the decisions to make when using MLPs, we have to decide the number of units in the hidden layer. This has close connection with the problem of generalization. We review this problem below, and the points raised there should be considered when determining the number of hidden units.

The Procedure of Learning for MLPs

- It is necessary to explore the range of the input signals, and make sure by scaling or other methods that there are no special cases where the input jumps into extreme values.
- The initial values of the connection weights are set to small random values in the vicinity of 0, such as in the range of $[-0.1, 0.1]$. If the connection weights are large, the units work in their saturation regions, $|f'(s)|$ becomes small, and learning becomes difficult. Furthermore, due to the structure of the MLP, hidden units are symmetrical. To break this symmetry, it is necessary to have random initial values for the connection weights.
- The error function G of the MLP is not convex, and generally has multiple optima. With an optimization based on the gradient method, finding a global optimum cannot be guaranteed. In practice, we can restart the procedure from different random initial values to cope with this.

- The error function G will not necessarily have 0 as the minimum. When the MLP cannot reconstruct the teacher signal, there is a positive remainder; this situation can occur for two reasons. One possibility is that the number of degrees of freedom of the MLP is too low. The other is that there is a contradiction or noise in the teacher signals that were supplied. It is necessary to review the situation to find out the real cause.

- Learning stops when there is no advantage in continuing. In practical terms, this is detected by seeing that the error G stays above some value, or the norm of the gradient $\nabla_{\boldsymbol{w}} G$ becomes close to 0. However, in the online case or in the case of the moment method, these values fluctuate, and the termination condition has to be determined by trial and error. Another method is to prepare a test data set, and test the network periodically after a certain number of learning steps; when there is no improvement with the test data, the learning can stop.

- The learning methods introduced so far are suitable for small-scale applications. However, they require trial-and-error adjustments in setting the learning coefficient and the termination conditions. For large-scale applications of the MLP, it is necessary to improve the computational efficiency and the parameter-setting process. We introduce more advanced learning methods below that should be considered for such applications.

- It is recommended to evaluate the MLP after learning by using test data separately from the training data. For this reason, it is necessary to reserve some data from the beginning for test purposes, and separate them from the training data.

10.3.4 Using More Refined Optimization Methods

The learning methods introduced in Section 10.3 are not quite satisfactory as optimization methods. As a result, 1) the convergence is slow and learning takes a long time; 2) it is necessary to do some experimenting to find a suitable learning coefficient; *etc.* In actual use of MLPs, we often need to repeat the learning in order to investigate closely the results, or to adjust parameters like the number of hidden units; so the above effects are a hindrance for researchers and practitioners.

Thus we can consider introducing more advanced optimization methods [71], however, in such a case some special conditions of the MLP should be taken into account.

- In many nonlinear optimization procedures, the following rules are applied repetitively:
  1. a rule to determine the direction to proceed from the current position,
  2. and a rule that determines the step length.

  For the latter rule, the so-called *linear search* [72], that is, optimization along a given direction, is used often. Nonlinear optimization in a strict sense requires the calculation of the gradient by batch-type learning, which

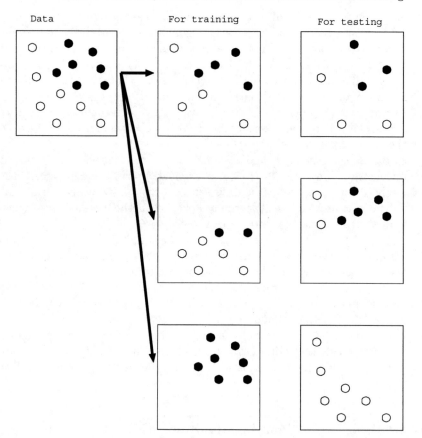

Fig. 10.8. Separation of data: good separation and bad separation

in the case of a large training data set will take a long time for each learning step. This indicates that we should try to find a way to perform as few linear search steps as possible.

- The number of connection weights in MLPs can easily be on the order of several hundred. Such methods that require the storage of second partial derivative matrices, like the Hesse matrix in the case of the Newton–Raphson method, are difficult to implement.

With the above considerations, a method that can be applied is the **conjugate gradient method** [73]. In the case of a quadratic objective function, this method computes the so-called *conjugate direction* vector set, and if linear search is used, it converges with just as few direction settings as the dimension of the objective function.

Another stable and fast method is the **quasi-Newton method**. However, it requires a storage space proportional to the square of the problem dimensionality, so it is difficult to use with the MLP. A method proposed by Saito *et al.* [74] modifies the quasi-Newton method by truncating the calculation of the BFGS equation, thus reducing the storage requirements; good results are reported.

10.4 Learning and Generalization by MLPs

10.4.1 Learning and Generalization

When some capability is obtained by learning, it is important to know what the answer is to the question of **generalization:**

"Can the machine, that succeeded in learning some given training data, also produce suitable responses for other data?"

We need to establish some conditions before we can speak about generalization:

1. It is necessary that some kind of hidden order be present in the target system, so that we can reasonably expect generalization to be possible. For instance, if we consider the mapping between names and phone numbers, in such learning problems there is no reason to expect to find any order, so there is no point in talking about generalization.
2. On the other hand, we need an ability on the side of the learning machine to replicate the order in the target system, without too large a degree of freedom. As an extreme example, just by memorizing the given training samples the training data can be easily learned, but there is no reason to expect any generalization.

10.4.2 Generalization in the Case of MLPs

In the case of MLPs, the learning takes place by the adjustment of the connection weights, under the restrictions of the network structure. This results in the acquisition of the hidden structure of the data, and thus in generalization. Thus we can say that

- For generalization, we should restrict the degrees of freedom (the number of hidden units and the number of connection weights) of the MLP.
- If the degree of freedom is too low, the training data cannot be reproduced;
- but if it is too high, learning becomes too easy, and we cannot guarantee suitable responses for such data that was not used in training.

10.4.3 Testing MLPs

As a basic method for ensuring generalization, we use the **testing of the MLP**.

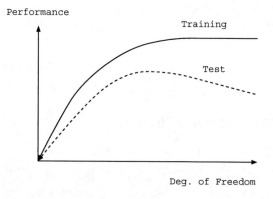

Fig. 10.9. Degrees of freedom and generalization of an MLP

- First, we separate the data into training and testing data sets.
- Next, we learn the network weights using the training data, and evaluate the trained MLP using the testing data.
- As we increase the degrees of freedom, *e.g.* the number of hidden units, the adaption of the MLP to the training data will continue to improve. On the other hand, adaption to the testing data will also increase while the degree of freedom is sufficiently low, but after a point it will start to decrease.
- Upon finding the point where adaption to the testing data starts to worsen, we can use it as a good setting for the MLP.

When there are only a few training data points, the method of cross validation [70, 75] can be used to increase the data that is available for training. Another method for improving the generalization is based on trying to reduce not only the error but also the number of nonzero weights during learning. There are also many other methods; for a review see [75].

10.4.4 Learning by Direct Optimization

For the learning of MLPs, we have assumed the existence of a training signal. However, in some applications such an explicit teacher signal cannot be obtained. For example, in a system with an embedded MLP where we can access only the overall performance value, we might want to optimize that by the adjustment of the MLP connection weights[4]. In such cases the gradient information is not easily obtained, and direct optimization methods are used.

10.4.5 Forward-Backward Modeling

Let us consider a situation where there is an unknown system connected behind the MLP, and we have its output as our teacher signal. This is a typical

---

[4] Optimal control, or reinforcement learning are such cases

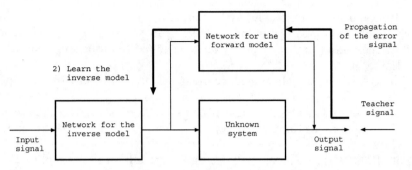

Fig. 10.10. Forward-Backward Modeling

situation in control problems. For the error backpropagation algorithm, we need the partial derivatives of the output of the unknown system, with respect to the input; *i.e.* the Jacobian matrix. However, usually we cannot easily obtain the Jacobian for an unknown system. In such cases, a technique used with MLPs is the *forward-backward modeling* [76] (see Figure 10.10). With this method, first we connect a MLP (the *forward model*) parallel with the unknown system, and train it to learn the input–output relation. Next, using the learned forward model, we determine approximately the Jacobian, and train the front MLP (the *backward model*). In control problems, in contrast with the forward model that learns the "Input → Output" relation, the backward model represents the relation of "Required output → The input needed to generate it", which is where this term comes from.

10.4.6 Learning with Powell's Conjugate Direction Method

If we consider the learning of the connection weights of the MLP as an optimization problem, it is not really necessary to use derivative information like the gradient. In situations where such information like the gradient is difficult to obtain, we can consider the use of direct optimization methods, which need only the objective function. One effective direct optimization method is *Powell's conjugate direction method*[72]. This method is similar to the conjugate gradient method introduced earlier, in that here also we generate a sequence of conjugate directions with respect to a quadratic function; however, it differs in that we do not use gradient information. Nakanishi *et al.* [77] have obtained good results for the learning of neural networks for control, using Powell's method.

10.4.7 Learning by Genetic Algorithms

Recently we are witnessing a rapid development of *genetic algorithms (GA)* as optimization methods. These are direct optimization methods, capable of global search, not requiring many assumptions about the structure of the problem, and useful in dealing with many kinds of objective functions. We have to note, however, that standard GA methods use binary representation of the genes, and they are generally not well suited to optimizing continuous functions.

For GA with real-valued gene representation, Ono *et al.* [78] have proposed the new *uniform normal distribution crossover (UNDX)* method, and reported about its good searching capabilities. This crossover method is capable of global optimization; moreover, it can deal with nonseparable objective functions, where there is some relation between the components. For the learning of MLPs, it is recommended to consider such advanced GA methods.

10.5 Reinforcement Learning

Let us consider a system operating in a given environment, which receives scalar-valued rewards or penalties as the evaluation (value) of its performance. In such a case, the process of maximizing its value by the mechanical learning of its behavior, is called *reinforcement learning*, and the value is called the *reinforcement signal*.

When considering such learning processes, some problems arise:

1. There can be a temporal delay between actions and evaluations, and we need to deal with this.
2. We need to find effective methods to search for the optimal behavior.
3. When making decisions, we need to deal with incomplete information about the environment.
4. We have to take into account the difference between simulations and reality.

In the following, we focus on 1 and 2 above, and as a preparation for introducing the reinforcement learning method called *Q-learning*, we review Markov decision processes and a solution method called *value iteration*.

10.5.1 Markov Decision Processes

The most important problem to solve in reinforcement learning is dealing with the temporal delays in evaluation signals. In other words, when a sequence of actions are made, and as a result evaluation signals are obtained, we have to trace back the relation between these signals and earlier actions. One framework for this problem is the Markov decision problem (MDP).

Let us consider an environment where

- a system is taking actions at the discrete times $t = 1, 2, 3 \cdots$;
- that system is in one of the *states* $x_i \in X$ ($|X|$ is finite);

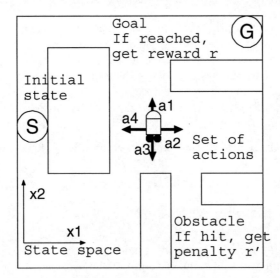

Fig. 10.11. Reinforcement Learning

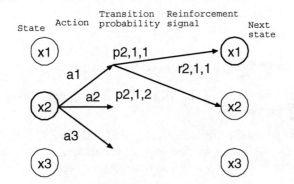

Fig. 10.12. A Markov decision process

- in each state, one of the *actions* $a_j \in A$ ($|A|$ is finite) is selected according to the state;
- when action a_j is selected, the system goes from the current state x_i to the new state $x_k \in X$ with probability p_{ijk}, and at the same time
- it receives the reinforcement signal $r_{ijk} \in R$.

We define the performance of the system as the *expectation of the discounted cumulative reinforcement signal*

$$V = \mathrm{E}\left(r(1) + \gamma^1 r(2) + \gamma^2 r(3) + \cdots\right) = \mathrm{E}\left(\sum_{t=1}^{\infty} \gamma^{(t-1)} r(t)\right) \qquad (10.21)$$

Here the constant $0 < \gamma < 1$ is called the discounting factor, as the one-step future reinforcement signal is discounted by γ when used as the value of the present. By introducing the discount factor, the value V is bounded when all r are bounded, even with an infinite number of time steps.

We denote by $r(t) \in R$ the temporal sequence of the reinforcement signals r_{ijk} received by the system. E() is the expectation w.r.t. the transition probabilities.

We call the function $\mu : X \to A$ that determines the action a_j for state x_i the *policy*. The Markov decision problem (MDP) is defined as follows:

MDP: find the policy μ^*, that maximizes the value V for a given initial state $x(0) \in X$.

That is,

$$\text{find} \quad \mu^* \quad \text{such that} \quad \mu^* = \arg\max_{\mu} V(x(0), \mu) \tag{10.22}$$

By formalizing the MDP, we have made clear the connection between the sequence of actions and the sequence of reinforcement values, and the solution of a reinforcement learning problem can be stated as the solution of an MDP. However, we should note that in the MDP formulation, there is no ambiguity in our knowledge about the environment, thus this is in a sense an idealized model.

10.5.2 Dynamic Programming (DP)

The basis for the solution of the MDP is dynamic programming. The idea of dynamic programming is

Principle of optimality: Assuming that future actions are optimal, and using their value in choosing a current action that is optimal according to that, the result is necessarily optimal.

This principle is used in many fields, *e.g.* in optimal control theory, or the computation of the shortest path in a directed graph *etc.*, and it is in general the principle behind the solution of multistep optimal decision problems. In the case of a finite number of time steps, dynamic programming can be applied starting at the final step and working backward. However, the MDP was defined earlier having an infinite number of steps, so we need some tricks to find a dynamic programming solution.

According to the dynamic programming method, we proceed thus:[5]

- Let the system be in state x_i, and we have to decide the action a_j.
- The state will make a probabilistic transition to x_k.
- Now let us assume that all future actions starting from x_k are decided according to the optimal policy μ^*. *
- Then we define the value of state x_k as $V^*(x_k)$.

---

[5] We might want to protest that we don't know this; but let us be optimistic and think only about the present, using this as a working hypothesis.

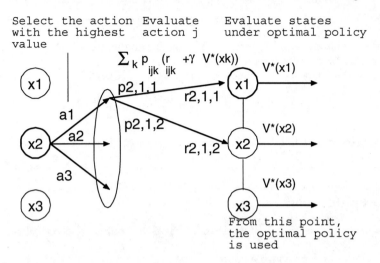

Fig. 10.13. Dynamic programming

Based on this hypothesis, the optimal decision in state x_i is the decision $a_j \in A$ that maximizes the expected value

$$\sum_k p_{ijk}(r_{ijk} + \gamma V^*(x_k)) \tag{10.23}$$

Therefore the optimal value V^* satisfies

$$V^*(x_i) = \max_{a_j} \sum_k p_{ijk}(r_{ijk} + \gamma V^*(x_k)) \tag{10.24}$$

Unfortunately Equation (10.24) has V^* on both sides, and also there is an operator for finding the maximum on the right-hand side, so it cannot be solved as it stands.

There are several algorithms for solving this; we are going to introduce one of them, the *value iteration* method, which has deep connections with the most famous reinforcement learning algorithm, *Q-learning*.

10.5.3 The Value Iteration Method

The value iteration method considers Equation (10.24) as an asymptotic equation for V^*, and obtains V^* as the converged value of an iteration[6]. In other words,

---

[6] With some assumptions it is possible to prove convergence.

Algorithm 10.1 Outline of the Value Iteration Algorithm for DP

1: set the initial value $V^0(x_i)$, $x_i \in X$ to some suitable value,
2: set the iteration counter as $l = 0$
3: from the following recurrence equation, set $V^l(x_i)$, $x_i \in X$,

$$V^{l+1}(x_i) = \max_{a_j} \sum_k p_{ijk}(r_{ijk} + \gamma V^l(x_k)), \quad \text{for} \quad x_i \in X \qquad (10.25)$$

4: stop the iteration if the value V^l has sufficiently converged; otherwise set $l = l+1$ and go to step 3.

10.5.4 Q-learning

Problems with Value Iteration (VI) as a Learning Procedure

The VI method provides a solution when all the parameters of the MDP are known. However, when dealing with reinforcement learning problems, in general we have to assume that there are unknown system parameters (like, *e.g.*, the transition probability matrix p_{ijk})[7]. Q-learning is a method for finding the optimal policy μ^* by trial and error, without requiring the explicit knowledge of the transition probability matrix p_{ijk} or the reinforcement signals r_{ijk}.

The Q Value

In Q-learning, we define the Q value as the function (table) $Q(x_i, a_j)$ of the state and action values x_i and a_j. $Q(x_i, a_j)$ has the meaning of

> The expected value, when in state x_i we take an (arbitrary) action a_j, and after that the system follows an optimal policy.

At convergence, the following relation will be satisfied by the optimal values V^* and the Q values at that time (denoted by Q^*):

$$Q^*(x_i, a_j) = \sum_k p_{ijk} \left(r_{ijk} + \gamma V^*(x_k) \right)$$

$$= \sum_k p_{ijk} \left(r_{ijk} + \gamma \max_{j'} Q^*(x_k, a'_j) \right)$$

$$V^*(x_i) = \max_{a_j} Q^*(x_i, a_j) \qquad (10.26)$$

Action Selection Rule

If we don't know the transition probability matrix p_{ijk} and the reinforcement signals r_{ijk}, we need to execute all valid actions (for a sufficient number of times) in each state, in order to be able to find the optimal policy.

---

[7] In fact, there would be no point in talking about "learning" otherwise.

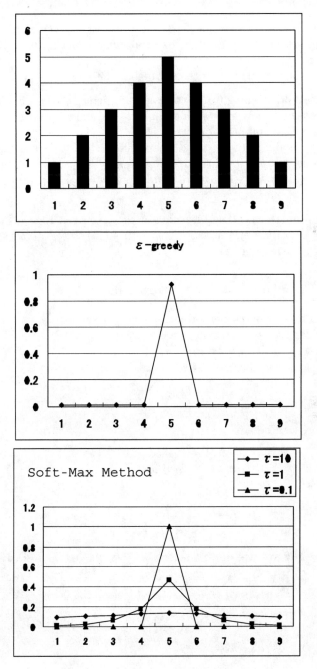

Fig. 10.14. Action selection rule

On the other hand, to make the most use of a given finite number of learning opportunities, we would like to concentrate on executing actions with higher expected rewards.

Some action selection rules that formalize the above are

ϵ-Greedy method : taking $0 < \epsilon < 1$ as a parameter, with probability $(1 - \epsilon)$ we select the action with the highest Q value, and with probability ϵ other actions with equal probability.

Soft-Max method : with $0 < \tau$ as a parameter, we select an action with probability proportional to $\exp(Q(x_i, a_j)/\tau)$[8]. If τ is small, actions with large Q values are selected with preference; for large τ values, all states are selected equally.

We should note that the Q-learning method, to be introduced later, poses only a mild condition on the action selection: in all states, all actions have to be selected sufficiently often.

Iterative Calculation of the Average Value

We want to make the VI method into a learning procedure; as a preparation let us consider the computation of the average of n data values. The average of the data $x_1, x_2, \cdots x_n$ is given by

$$\bar{x} = \frac{1}{n} \sum_{i=1}^{n} x_i \tag{10.27}$$

which in turn can be written as

$$\bar{x} = \frac{1}{n} \left(\sum_{i=1}^{n-1} x_i + x_n \right)$$
$$= \frac{n-1}{n} \left(\frac{1}{n-1} \sum_{i=1}^{n-1} x_i \right) + \frac{1}{n} x_n \tag{10.28}$$

If we denote the average up to t points by $\bar{x}(t)$, we get the iterative formula

$$\bar{x}(t+1) = \left(\frac{t}{t+1} \right) \bar{x}(t) + \left(\frac{1}{t+1} \right) x(t+1)$$
$$= (1 - \alpha(t+1))\bar{x}(t) + \alpha(t+1)x(t+1) \tag{10.29}$$

Here $\alpha(t+1) = 1/(t+1)$.

---

[8] By analogy with the Boltzmann distribution in statistical mechanics, τ is called the *temperature*.

Learning of the Q Value

Let us consider a procedure where at time t, with the system in state $x(t) = x_i$, we use *e.g.* the ε-greedy method to determine the action $a(t) = a_j$, and the system makes a (probabilistic) transition to state $x(t + 1) = x_k$, and we receive a reinforcement signal $r(t)$. From the VI method, let us consider how to execute this by learning, while p_{ijk}, r_{ijk} are unknown.

First, the VI updating rule of Equation (10.25) can be written as

$$Q^{t+1}(x_i, a_j) = \sum_k p_{ijk}(r_{ijk} + \gamma V^t(x_k)) \tag{10.30}$$

$$V^{t+1}(x_i) = \max_{a_j} Q^{t+1}(x_i, a_j) \tag{10.31}$$

Furthermore, by substituting (the one step earlier) Equation (10.31) into Equation (10.30), we get

$$Q^{t+1}(x_i, a_j) = \sum_k p_{ijk}(r_{ijk} + \gamma \max_{a_l} Q^t(x_k, a_l)) \tag{10.32}$$

Instead of calculating the expectation from p_{ijk}, we use the actually observed transitions $x(t) \to x(t + 1)$ in an iterative procedure to calculate Q as

$$Q^{t+1}(x(t), a(t)) = (1 - \alpha)Q^t(x(t), a(t)) + \alpha(r(t) + \gamma V(x(t + 1)))$$
$$V^t(x(t + 1)) = \max_{a_j} Q^t(x(t + 1), a_j) \tag{10.33}$$

Equivalently, we can write it as

$$Q^{t+1}(x(t), a(t)) =$$
$$(1 - \alpha)Q^t(x(t), a(t)) + \alpha(r(t) + \gamma \max_{a_l} Q^t(x(t + 1), a_l)) \tag{10.34}$$

Here $0 < \alpha < 1$ is a parameter called the *learning coefficient*, which is needed to calculate the averages under the uncertainties of the transitions and reinforcement signals. As we can guess from the discussion about the average calculation, in order to ensure the convergence of the Q value, we have to decrease α gradually.

11

Genetic Algorithms for Control-system Optimization

11.1 Stochastic Approach to Optimization

An effective approach to various problems in information processing is

- formulate the problem as an optimization task,
- then solve it by numerical methods.

For example, to restore an image that is distorted by noise, we can formulate the problem by defining an evaluation function that encompasses the requirements of "conforming to the observed data" and "conforming to image-specific properties like the continuity of the intensity across neighboring pixels"; then optimize this function with respect to the restored image data. The advantage of this approach is that by formulating the task as an optimization problem, it becomes well defined; Also, we can deal separately with the formulation and the solution phases, and for the solution we can develop or use generic optimization methods.

The process of formulating a task as an optimization problem depends on the target problem area; on the other hand, the solution method of the resulting optimization problem depends on whether the decision variable is discrete or continuous, whether there are constraints or not, *etc.* One method for getting analytical ("exact") solutions is the *Linear Programming (LP)* method. There is a rich literature about LP, and with, *e.g.* the simplex method or with interior-point methods we can solve problems with thousands or tens of thousands of variables. However, methods offering analytical solutions are available only for specific classes of problems, and therefore their applications are also restricted.

Some classes of problems where we cannot expect much from analytical methods are the combinatorial optimization area, where many problems are NP-hard; or the optimization of nonlinear functions, where often there are many local optima. On the other hand, from the standpoint of practical applications, we often do not require exact solutions, but will be satisfied instead with good approximate solutions; so there is a need for developing methods of approximate solution. Stochastic search is one class of such approximate solution methods. Representative examples of stochastic search include simulated annealing, Hopfield's method, tabu search *etc.* [79]. In this section, we introduce genetic algorithms (GA), which also belong to this class, from the point of view of stochastic information processing.

11.2 Genetic Algorithm

The genetic algorithm (GA), as its name indicates, is based on Darwin's theory of natural selection, and performs optimization and other tasks by modeling the process of genetic evolution of biological organisms. GA itself had been proposed fairly early by Holland and his group, but it became popular only after the 1980s, when it became quickly accepted worldwide.

In GA, for the problem we want to solve, we define each candidate solution as an *individual* like that of an organism, collect such individuals into a *population* (a collection of individuals), and we perform the operations of crossover, mutation, selection, *etc.*, as steps in the search. The basic structure of GA is the following:

Algorithm 11.1 Outline of the structure of GA

1: Create the initial population. Usually we generate individuals randomly.
2: **Crossover** By combining two or more individuals, we create new ones.
3: **Mutation** By random perturbation on an individual, we create a new one.
4: **Selection** We evaluate the fitness of each individual, according to the objective function of the optimization task, and remove those with low fitness values.
5: The search is terminated by some appropriate criteria, or continued from Step 2.

GA has some special features: one is that it uses a population, *i.e.* a set of solutions, for its search; another is that it uses the crossover operation to combine several solutions into a new one. Other stochastic search algorithms, such as simulated annealing, usually employ a single candidate solution, and thus usually they need to define some sort of neighborhood of the current solution to be able to generate a new one.

Such features of GA have contributed to making the design of solution methods more free, and allowing their application in a wide area; however, at the same time they have made the theoretical analysis difficult, and finding efficient methods is also not easy. In fact, even during the worldwide research boom in this area, it took quite a long time for GA to develop into an efficient optimization method.

In the following, we will introduce two application areas of GA: combinatorial optimization, which operates on a discrete domain; and nonlinear optimization, which works on a continuous domain. Besides these examples, GA is widely used, *e.g.*

- in the case of multiple objective functions, where the tradeoff frontier is explored [80],
- for evolving programs or other non-numerical structures by the methods of genetic programming [81],
- for optimizing by running large-scale simulations for evaluating the objective function, or optimizing by experiments [82]

and in many other areas of research.

11.2.1 Combinatorial Optimization with GA

GA is an effective approximate optimization method for combinatorial optimization problems; the key to its application is the design of the crossover operation. A combinatorial optimization problem that is easy to visualize is the traveling salesman problem (TSP) , where we need to find the shortest tour (closed path) going through a set of given spatial points (cities); in the crossover, we need to generate a new path from two (or more) existing paths. In this case, the advantage of searching by GA is realized in the combination of two different paths into a new one. However, as is illustrated in Fig. 11.1, the combination of the tours of the two parents is in itself subject to a combinatorial explosion of possibilities; on the other hand, as the search progresses, the combined tours are apt to degrade from the parent tours, and it becomes more and more difficult to combine the good parts of the parents, even if such a possibility does exist in principle.

The first textbook of GA research [83] introduced a commonly used method of crossover for TSP: the tours are expressed as sequences, and two tours are combined in a way to keep the result a proper sequence. However, with this operation it is easy to destroy such structures of the tours

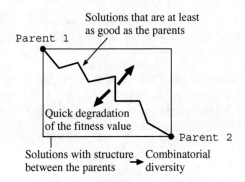

Fig. 11.1. Crossover in a combinatorial optimization problem

like "short branches", that are important to preserve for obtaining good solutions; and therefore even problems with tens of cities are difficult to solve
with this method. As the research progressed, later there came proposals
for other crossover methods better suited to TSP, with attention paid to
branches and subtours: the subtour exchange crossover (SXX) or the branch
exchange crossover (EXX). Also, there has been research on genetic local
search, which performs a local search in the vicinity of individuals generated by
the crossover, and the search proceeds by using such improved individuals. As
a result, it became possible to solve problems with hundreds of cities [84, 85].

However, although SXX or EXX are good at inheriting branches from
the parents, this property in turn impairs their ability to generate diverse
solutions. Nagata and Kobayashi [86] have proposed as a solution the branch
construction crossover (EAX), which combines the properties of inheriting
branches from the parents, while retaining the diversity of the offspring; this
has made it possible to solve problems with thousands of cities. This is a level
that is difficult to reach for other competing approximate solution methods.

We can see that the key to the effective application of GA is the proper
design of the crossover operation. In an easily visualized problem like TSP, we
can rely on intuition to design a domain-specific crossover method; however,
it would be desirable to find some more generic design method. As an answer
to such requirements, a proposed methodology starts not from directly combining two solutions, but instead defining the comparatively easier concept
of neighborhood (distance) for a single solution, and approaching from one
solution in the direction of another. Yamada and Nakano [87] have added to
this approach some concepts from simulated annealing, and developed the
method of multi-step crossover fusion (MSXF); Ikeda [88] further simplified
this method without significantly reducing its search capability, and proposed
the method of "crossover by picking the best parts" that also combines the
selection operation:

Algorithm 11.2 Crossover for GA

1: Let the parents be p_1, p_2. Let the first search point be $x_1 = p_1$, with $k = 1$.
2: In the neighborhood of x_k, denote the set that approaches p_2 as $\mathcal{N}_{\to p_2}(x_k)$; the
   maximum allowed search count here is μ.
3: The element in $\mathcal{N}_{\to p_2}(x_k)$ with the best evaluation becomes the next search
   point x_{k+1}. If the search point has reached p_2, or the step count k has reached
   $k_{\max}$, go to 4. Otherwise, set the step count to $k = k + 1$, and go to 2.
4: From the search points $x_1, \cdots, x_{k_{\max}}$, select the best one and replace in the
   population p_1 with it.

11.2.2 Nonlinear Optimization with GA

A further important application of GA is the optimization of nonlinear functions in the continuous domain. When the objective function is multipeaked with many local optima, approximate stochastic global optimization methods like GA are effective; such problems are often encountered in engineering. In the beginning, standard GA methods, which are designed for discrete problems, were used directly to solve such problems, by discretizing the continuous problem domain, in the so-called "bitmapped GA" approach. However, such approaches destroy the continuity inherent in the problem, thus making the solution more difficult. Furthermore, in nonlinear optimization, often we have to deal with problems where the decision variables have to be set in some correlated way, making the problem "nonseparable". Bitmapped GA, or even continuous-valued methods with crossover procedures acting independently upon the decision variable components, are no match for problems with severe nonseparability.

Considering the above, there are crossover operations with good search capabilities, that are based on generating the offspring from the linear combination of the parents. As representative examples, we introduce the UNDX-m method that is the extension of the uniform normal distribution crossover (UNDX) of Ono *et al.* [78] to multiple parents; and the Simplex Crossover (SPX) by Tsutsui *et al.* [89][1]:

UNDX-m

In UNDX-m, we generate an offspring according to a normal distribution around the center of gravity of the parents:

SPX

In SPX, the offsprings are generated according to a uniform distribution inside the simplex (convex polytope) of the parent individuals:

A recommended value for α is

$$\alpha = \sqrt{n+2}. \tag{11.7}$$

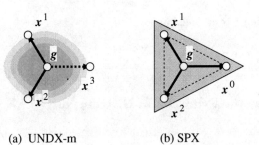

    (a) UNDX-m   (b) SPX

Fig. 11.2. UNDX-m SPX

---

[1] For a more detailed exposition see [75, 78, 85].

Algorithm 11.3 The UNDX-m algorithm

1: Select randomly $m + 1$ individuals from the parent population $x^1, ..., x^{m+1}$. m is on the order of 1–5.
2: Define the centroid of the parents as $g = \frac{1}{m+1} \sum_i x^i$, and denote the vectors from the centroid g to each parent x^i by $d^i = x^i - g$.
3: Select one more parent x^{m+2} randomly from the population.
4: Set D to the length of the component of the vector $d^{m+2} = x^{m+2} - g$, going from the parent x^{m+2} to the centroid g, that is perpendicular to the vectors $d^1, \cdots, d^m$.
5: $e^1, \ldots, e^{n-m}$ is the orthonormal base of the subspace orthogonal to $d^1, \cdots, d^m$.
6: The offspring x^c is generated as:

$$x^c = g + \sum_{k=1}^{m} w_k d^k + \sum_{k=1}^{n-m} v_k D e^k \tag{11.1}$$

Here w_k and v_k are random numbers from the distributions of $N(0, \sigma_\xi^2)$ and $N(0, \sigma_\eta^2)$. Also, σ_ξ and σ_η are parameters, with the recommended values of

$$\sigma_\xi = \frac{1}{\sqrt{m}}, \quad \sigma_\eta = \frac{0.35}{\sqrt{n-m}} \tag{11.2}$$

Algorithm 11.4 The SPX algorithm

1: Select $n + 1$ parents $x^0, ..., x^n$ randomly from the population.
2: Call the centroid by $g = \frac{1}{n+1} \sum_i x^i$.
3: Let $c^0 = 0$, $p^0 = g + \alpha(x^0 - g)$.
4: For c^k, p^k $k = 1, \cdots, n$, evaluate the formula

$$p^k = g + \alpha(x^k - g) \tag{11.3}$$
$$c^k = r_{k-1}(p^{k-1} - p^k + c^{k-1}) \tag{11.4}$$

Here α is a positive constant, r_k is calculated from a uniform random number in $[0, 1]$ as

$$r_k = (u(0, 1))^{\frac{1}{k+1}} \tag{11.5}$$

5: The offspring c is obtained as

$$c = p^n + c^n \tag{11.6}$$

11.2.3 GA as the Evolution of Distributions

There is no shortage of theoretical research about GA, but we cannot say that this research has definitely contributed to the development of GA methods with excellent search capabilities. The most important characteristic of GA that distinguishes it from other stochastic search algorithms is the crossover operation, used in the generation of candidate solutions. The facts that crossover is an operation on multiple individuals; or that if the diversity is lost in the population, crossover will no longer work, no matter if the solutions are good or bad; are possible reasons for this.

However, if instead of viewing crossover in a microscopic way, that is, as an operation for generating offspring from parents, we view it as the global evolution of the population of GA, it becomes easy to visualize crossovers that lead to good search capability.

Let us assume that the search has progressed to some extent, and we have a population with good evaluation values (Fig. 11.3a). Now suppose that crossover has acted on this population, and a new population had been generated.

If the crossover is such that it results in search points in larger areas than the area of the already obtained good-quality population (Fig. 11.3d), that means that we are searching again in such unpromising places that were already abandoned by the search process, which is useless. On the other hand, if the crossover is such that new search points are generated only in some specific areas of the parent population (Fig. 11.3b), that means that searching some areas had been abandoned without good reason, thus we cannot expect the process to be a thorough search. Thus the desirable crossover is such that it does not distort significantly the distribution of the population (Fig. 11.3c).

In a discrete search space, it is not evident how the concept of neighborhood can be introduced, and it would depend on the coding of the solution, thus the above discussion is difficult; however, in nonlinear optimization with a continuous search space, the distribution of the population can be treated *e.g.* by statistical methods. In the case of the UNDX-m or the SPX methods introduced above, in order not to modify significantly the characteristics of the population, the recommended parameters are selected in such a way so as

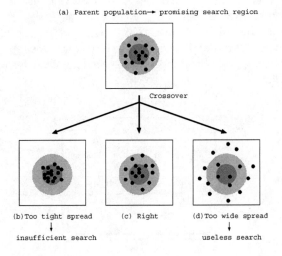

Fig. 11.3. Evolution of the population distribution according to crossover

to preserve the mean, variance, and covariance matrix of the population; in fact, benchmark tests have proven their capabilities.

This ability of GA to adaptively sample the search space, which is due to the crossover operation that combines multiple parents, is a great advantage of GA. In contrast, search methods for continuous spaces that use random perturbation of the search points, that is, mutation, need to determine the extent of the perturbation by some method, which can make the algorithms complicated.

11.2.4 GA and Estimation of Distributions Algorithms

A family of algorithms that originated in GA is the estimation of distribution algorithms (EDA). In EDA, the following steps are used:

Algorithm 11.5 Outline of the estimation of distribution algorithm for GA

1: Generate an initial population randomly.
2: From the sample population, select n solutions with good objective function values, where n is the basis dimension.
3: If a termination condition is met, stop. Otherwise, go to Step 4.
4: Generate a distribution model that describes the selected subpopulation.
5: Resample according to this model.

As seen in Fig. 11.4, in contrast to GA, which generates the offspring from the parents by crossover and mutation, in EDA, new search points are generated by the operations of "estimation of model distribution" and "sampling by the model". The simplest model uses independent distributions for each component; more advanced models use correlations among the components, or more generally Bayesian networks (Kurahashi and Terano [90] have a more detailed explanatory article). For applications to continuous spaces, research is being done on using probabilistic principal component analysis, or Helmholtz machines [91, 92].

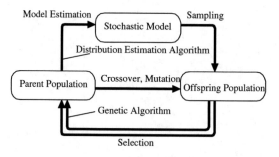

Fig. 11.4. GA and EDA

From the above, if we contrast GA and EDA, we can say that they have similar goals but employ different methods. To generate new solution samples with similar distribution to the current population, GA uses the crossover of individual parents, in what could be called a microscopic approach; while EDA introduces an explicit model for the distribution, in a macroscopic approach. This structural difference causes the former approach to be simpler, but at the same time to require some kind of "art" in the design. The latter approach has a more complicated structure, but the meaning is explicit. In terms of their advantages and disadvantages as optimization methods, there is more research to be done; but these two approaches have already provided interesting insights, both from the point of view of probabilistic models, as well as the theory of GA. As a further advance, there is some interesting research about regarding the crossover in GA as a model of probabilistic distributions [93].

11.3 Optimization of Uncertain Fitness Functions by Genetic Algorithms

11.3.1 Introduction to GA for Optimization with Uncertainty

Genetic algorithms (GA) are computational methods that are modeled on the adaptive evolution of living organisms [84]; in contrast to their simple structure, they are now used in much wider areas than only optimization. Especially in recent years, as the capabilities of personal computers (PC) are rapidly improving at lower and lower costs, GA methods, which are well suited for PC clusters, are entering application areas that require high computing capacities. Some of the new application areas of GA, even without considering such direct applications like "artificial life", are the approximate solution of NP-hard combinatorial optimization problems; the solution of optimization problems with continuous-valued objective functions by real-valued GA; and developing GA methods for optimization problems with multiple objective functions. Genetic programming (GP), the study of a branch of GA that optimizes tree structures, programs, circuits *etc.* instead of parameters, is another active research area.

From the wide range of applications of GA, here we consider their use for the optimization of objective functions with uncertainty, in particular the topics of "optimization of noisy fitness functions", and "adaptation to changing environments". We give formal definitions for these problems, and describe the practical requirements from their application, then introduce the approaches from GA. Finally, we review the application of the methods proposed by the authors to these problems.

11.3.2 Optimization of Noisy Fitness Functions

For a fitness function $f(\boldsymbol{x}, \boldsymbol{\delta})^2$, which includes an uncertainty $\boldsymbol{\delta}$, let us consider its optimization

$$\min_{\boldsymbol{x}} f(\boldsymbol{x}, \boldsymbol{\delta}) \tag{11.8}$$

by the decision variable $\boldsymbol{x}$. In this formulation, the uncertainty $\boldsymbol{\delta}$ is left undefined. We next consider a more restricted version of the problem (11.8), the problem of "optimization of a fitness function including observation noise":

$$\min_{\boldsymbol{x}} \langle f(\boldsymbol{x}) + \delta \rangle_\delta. \tag{11.9}$$

We assume that the fitness function is the sum of an additive scalar observation noise δ and its true value. Also, we denote the expectation of the uncertainty δ by $\langle \rangle_\delta$, and assume that $\langle \delta \rangle_\delta = 0$.

A similar problem is when we require a solution that does not degrade significantly with changes in the environment, that is, a "robust optimum". There are several possible formulations, but in the GA literature, it is usual to consider a disturbance $\boldsymbol{\delta}$ in the input, and to seek to optimize the expectation of the performance [89, 94–96]:

$$\min_{\boldsymbol{x}} \langle f(\boldsymbol{x} + \boldsymbol{\delta}) \rangle_\delta \tag{11.10}$$

There are many similarities in the methods of solution for the above two problems, but we also need to be aware of the differences. In the case of the first one, in actual applications we have no access to the true fitness function $f(\boldsymbol{x})$, and we can use only the observed values $f(\boldsymbol{x}) + \delta$. Also, in this case the noise is single-dimensional. In contrast to this, in the second problem, the nominal fitness function $f(\boldsymbol{x})$ is known, and we can make *a priori* assumptions about the environmental changes $\boldsymbol{\delta}$; thus in principle the fitness function $\langle f(\boldsymbol{x} + \boldsymbol{\delta}) \rangle_\delta$ is computable. However, since generally the input is multidimensional, thus the evaluation of the expectation with respect to the disturbances $\boldsymbol{\delta}$ requires difficult integrations, it is customary to use approximations by random sampling.

11.3.3 Adaptation to Changing Environment

Another optimization with uncertainty is "adaptation to changing environment", which is the subject of many research projects in GA. It can be formalized similarly to the previous section:

$$\min_{\boldsymbol{x}_t} f(\boldsymbol{x}_t + \boldsymbol{\delta}_t), \quad t = 1, 2, \cdots \tag{11.11}$$

---

[2] In optimization theory, this is called the "objective function" or "value function", but here we defer to the custom of GA and call it the "fitness function". However, the following formalization is done according to optimization theory.

That is, as the environment changes with time t as δ_t, we want to change the decision variable x_t.

During the determination of the decision variable $\boldsymbol{x}_t$, we need to define what information is available about the function f and the environmental variations $\boldsymbol{\delta}$. For instance, if we can assume that the change $\boldsymbol{\delta}_t \to \boldsymbol{\delta}_{t+1}$ in the environment is small, we can find efficient methods to search for the optimal solution. Also, if we can estimate to some extent the variation δ_t, or if we know that some variation appears repetitively, these properties can be expected to be useful for constructing efficient optimization methods.

As can be seen from the above exposition, the difference between the two formulations lies in the relation between the variation $\boldsymbol{\delta}$ and the decisions $\boldsymbol{x}$. In practice, there can be cases with elements from both.

11.3.4 Discussion from the Application Side

In general, the optimization of systems with uncertainties that we have defined above is more difficult than the case of systems without uncertainties. Therefore it is important to consider the actual requirements from the applications' side, when developing methods of solution. We review two applications with the above in mind.

Online Optimization

One application is "optimization by experiments" [3], or "online optimization". Often we want to tune optimally the decision parameters of some equipment, but because of the complexity of the equipment or its working environment, we cannot construct sufficiently detailed simulation models for it. Or even if simulation were theoretically possible, it might not be economically feasible.

In such cases, we can resort to tuning the parameters experimentally, by actually operating the equipment; it is obviously desirable to automate this process and tune automatically the parameters during the experimentation. It is even more desirable for the equipment to be able to tune itself "online", while operating in its actual operating environment. The proliferation of microcomputers built in to various controlling and measuring equipment has contributed to this requirement; however, in such cases the system performance can only observed with noise, and the operating environment will also not be static, so adaptation to environmental variations will be needed.

---

[3] Similarly to GA, evolution strategies (ES) have also been developed as optimization methods based on the model of biological evolution [84], and in the beginning they were proposed as a framework for optimization by experimenting; thus in ES there have been both algorithmic and theoretical approaches at optimization of noisy fitness functions[97–101].

Optimization by Simulation

Another application is "optimization by simulation". In the case of large-scale and complex systems, often it is not feasible to determine theoretically the optimal setting of their tunable parameters; on the other hand, we might not be able to use experimenting either, because of costs, time constraints, safety concerns, *etc.* In such cases we might need to resort to tuning the parameters through simulation experiments.

In the past, limitations on the computing resources have made parameter tuning through simulation infeasible in many cases, allowing at most the comparison of a limited number of variants. However, recent developments both in the computing power of individual computers, and the increasing availability of PC clusters, are making repetitive simulations a realistic choice in many cases. However, for instance in the simulation of traffic systems the traffic demand will be generated by using random samples, and such randomness is found in many simulations; thus such applications will need optimization methods that are capable of dealing with noise.

Requirements of Applications

Upon reviewing these applications, we can find some requirements on the optimization methods. One of them is "avoiding experiments with extreme settings". In the case of experimental equipments, it goes without saying that experiments with extreme settings can be dangerous and should be avoided; but even with simulations, although physical dangers can be discounted, simulation experiments with extreme settings can cause unexpected phenomena, for which the simulator itself might not have been designed, and the results will not be reliable. These concerns, however, can be addressed to some extent by limiting the search space.

A more difficult requirement is to put limits on the number of evaluations. In experimental optimization, the time requirements, the lifetime of the equipment, or the workload on the experimenters can severely limit the number of times of evaluations. Similarly, in the simulation of large-scale complex systems, the simulation time is not negligible, and that will put a limit on the number of evaluations. In practice, we would need to estimate the allowed number of evaluations, and structure the problem and its solution method around that limitation.

11.3.5 Approach to Uncertain Optimization by GA

We consider GA an appropriate choice for solving optimization problems with uncertainty, because

- it is an optimization method that requires only the evaluation of the fitness function, and does not require gradients or other similar information, thus

it fits well the experimental or simulation framework, where only the fitness value is available;
- it can perform a global search in the required region;
- as it uses a population-based search, we can expect it to be a robust method.

In particular, there have been proposals for methods that are explicitly designed for optimization with uncertainty, and we survey them in the following.

11.3.6 GA for Optimizing a Fitness Function with Noise

We include methods proposed for robust optimization in the following, because of their common structure.

Using Standard GA

Since GA searches by a population, even if the fitness of individuals is noisy, this will be averaged out in the population, and it will have the property of optimizing the expectation of the fitness function [89]. Thus one possibility would be to use standard GA for optimization with noisy fitness functions. However, this would require rather large population sizes and a large number of evaluations, so for practical uses we need to search for more refined methods. Another possibility is a proposal in the ES framework by Markon *et al.* [102], of modifying the selection process, and accepting new individuals only if their fitness is better than the existing ones by at least a given threshold.

Resampling of the Fitness Value

A general method for reducing the noise in the fitness values is to generate multiple samples for each individual, and to use their average. Assuming independent noise, with N samples the variance will be reduced as $1/N$ (thus the standard deviation is reduced by $1/\sqrt{N}$). Fitzpatrick and Greffenstette [103] have proposed a GA method where each individual is resampled several times, and the GA proceeds by using the averages as fitness values. This method has the advantage of not requiring special assumptions on the fitness function, but it can require a high number of evaluations to achieve the necessary reduction in noise. There are several proposals on reducing or optimizing the number of samples. Aizawa and Wah [104] have considered the tradeoff between the population size and the sampling count. Branke [94] has proposed concentrating the resampling to the best individual and its vicinity; Stagge [105] proposed determining the sample count by t-significance test.

Using the Fitness Values of Other Individuals

As a method of further reducing the noise in the fitness values, instead of using only the sample values of each individuals, we can try to incorporate the sample values from other individuals. As a precondition, we need to make some assumptions about the fitness function.

One commonly used (either explicit or implicit) assumption is "near solutions have near true fitness values". As a practical example, Tamaki and coworkers [95], or Tanooka and coworkers [96] have proposed that the parent fitness value is inherited by the descendants. However, with this method, since the parents are the survivors of the selection process, this causes their fitness to have a bias, which is inherited by their offspring. Branke [94] has proposed using the fitness value of individuals located between the parent and the offspring. One of the authors and others [75, 106, 107] have proposed the "memory-based fitness evaluation GA" (MFEGA), which also belongs to this category, but it retains all the individuals and their fitness sample values that have been generated during the optimization process, thus achieving a higher accuracy in the estimation of the fitness function. We describe MFEGA in more detail in the next section. A similar method has been proposed by Yamamoto and coworkers [108, 109]; in that case, instead of improving the estimation of the fitness function, they use the estimated fitness directly for the evolution process, to speed up the evolution as compared to the process that uses the sampled values.

From the above, we can see that much research has been done for developing GA methods for noisy fitness functions, but from practical requirements, the limitation on the allowable function evaluation is rather strict. Thus dealing with noise by multiple sampling is not practical, and even if it requires making assumptions about the fitness function, active reuse of existing sample values becomes an attractive choice.

11.3.7 GA for Varying Environments

Next, we consider GA methods proposed for adapting to environmental variations. We can divide the strategies used to develop such methods into two main classes.

Methods for Retaining the Search Capability

In GA, the individuals will normally converge around the optimum, which increases the accuracy of the solution. On the other hand, if we want to be able to adapt to environmental changes, the location of the optimum will change, and GA will need to keep its search capability so that it can track it.

As the simplest method of dealing with variable environments, we can restart the GA when an environmental change is detected[110–113]. A more controlled method for retaining the search capability had been proposed by

Grefenstette [114]; he calls the method of injecting new individuals into the population "random immigrants". Cobb and coworkers [115, 116], on the assumption that environmental variations can be detected, proposed the method of "triggered hyper mutation", where upon detection of the change, the mutation rate is increased. In contrast to these methods, which use a forced supply of diversity to retain the search capability, Mori *et al.*'s [117] method of "thermo-dynamical GA" (TDGA) retains the diversity internally. In TDGA, the diversity of the individuals is explicitly included in the evaluation for selection as "entropy", and it is able to control the diversity according to the environmental variation.

The Strategy of Recording the Search Results

In such cases when the environmental variation is recurrent, that is, we can assume that an environment that has occurred once will appear again, an effective method is to remember and reuse the past adaptation results.

As a method of implementing this, there is a lot of research done on introducing the framework of diploidal models into GA, from the biological principle of diploidal structures, *i.e.* having two sets of corresponding chromosomes, found in many sexually reproducing organisms [118–122].

More generally, we can view this method as a way of introducing redundancy into the genetic expression, and some similar approaches are the structured gene model [123] or the directed evolutionary model [124]. However, in such methods with redundant genetic expressions, it is difficult to statistically process the recorded search results and their correlations, and also it is difficult to protect the recorded results from the effects of selection, crossover, and mutation.

As a solution to this, Mori *et al.* [125] have proposed the "immune algorithm (IA)", taking as a model the rapid immune response to such antigens that had been encountered in the past. In IA, the past search results are recorded and handled separately from the population that is used for search. We can view the IA as a way of effectively combining the concepts of GA and case-based modeling. By introducing into IA the methods of TDGA, for maintaining the diversity in the search population and in the recorded population, we obtain the "memory-based TDGA" [126]. As another approach, Trojanowski and Michalewicz [127] have proposed recording the parent individuals into an individual memory buffer, and using them when an environmental change is detected.

As seen from the above discussion, much research is published about developing GA methods to deal with environmental variations; however, we cannot say that they always take into account the practical requirements of the applications, and some of them tend to be rather theoretically oriented. In the work of the authors called *genetic algorithm using subpopulation (GASP)*, introduced in [128], which belongs to the class of "strategies of recording search results", the problem definition is based on the requirements of the actual applications, and to get a practical method, we take into account the prior and posterior estimates of the environment.

11.3.8 MFEGA and an Example of its Application

The basic principle of MFEGA [75, 106, 107] is to keep a record of all samples obtained during the search, and when a new search point is generated, we calculate a more precise estimate of the fitness value using both the sampled fitness value, and the recorded historical sample values, and use that in the search process.

For such estimation of the fitness function, it is necessary to introduce some kind of model for it; in MFEGA, we make the following very simple assumption about the noisy fitness function: "In the vicinity of the search point under consideration (in the following we call it simply the search point), the fitness values are randomly distributed around the true fitness value of the search point". We further assume that the variance x of this random distribution is proportional to the distance d from the search point, as shown in Fig. 11.5. That is,

$$f(h) \sim N(f(x), kd) \tag{11.12}$$

$$\delta \sim N(0, \sigma_E^2) \tag{11.13}$$

$$F(h) = f(h) + \delta \sim N(f(x), kd + \sigma_E^2) \tag{11.14}$$

Here $f(x)$ is the true fitness value at x, and k is a positive parameter. Also, we assume the randomness to follow a normal distribution.

According to this model, the true fitness function value $f(x)$ can be estimated by the maximum likelihood method as

$$\tilde{f}(x) = \frac{F(x) + \sum_{l=2}^{H} \frac{1}{(k'd_l + 1)} F(h_l)}{1 + \sum_{l=2}^{H} \frac{1}{(k'd_l + 1)}} \tag{11.15}$$

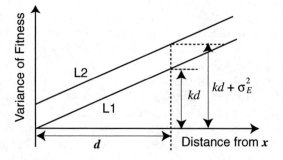

Fig. 11.5. Probability model of the fitness function

Here the H recorded search samples are $\boldsymbol{h}_l, l = 1, ..., H$, $F(\boldsymbol{h}_l)$ and $d_l, l = 1, 2, ..., H$ are the sampled fitness function values and the distances from the search point $\boldsymbol{x}$. Also, $k' = k/\sigma_E^2$.

In the actual algorithm of MFEGA, we simultaneously estimate the parameter k', and run GA with the estimated value $\tilde{f}(\boldsymbol{x})$. However, such an estimation has a tendency of supressing extrapolating search. Because of this, we eliminate during the selection and generation change the worst outlying fitness sample values[129].

As an example of online optimization, the authors have applied MFEGA to the optimization of the fuel control of an engine during acceleration [130]; also, as an example of simulation-based optimization, we have used it to optimize the control strategy for multicar elevators (MCE) [131, 132]. We introduce the latter application in Chapter 18.

Control System Optimization by Evolution Strategies and Particle Swarm Optimization

This chapter gives a comprehensive introduction into *evolution strategies* (ES) and *particle swarm optimization* (PSO). The former belong—together with *genetic algorithms* (GA) and *genetic programming* (GP)—to the class of *evolutionary algorithms*. Readers who are interested in the history of EA are referred to [133].

A stochastic approach has been used in Chapter 11 to introduce genetic algorithms. Here, we will complement this approach by presenting design principles and theoretical issues for evolution strategies. Particle swarm optimization has been included in this introduction, because it has gained much attention during the last decade and has been successfully applied to many real-world optimization problems. Differences and similarities of these algorithms will be discussed, too. The applicability of ES and PSO to control-system optimization will be demonstrated in Chapter 16, whereas the present chapter presents fundamental concepts.

12.1 Evolution Strategies

12.1.1 Framework of Evolution Strategies

Evolution strategies were developed in the 1960s as rules to generate experimental designs and to analyze experiments—surprisingly, they have not been developed to determine extrema of real-valued objective functions [134]. The first $(1+1)$-ES used binomially distributed mutations [135]. They have been replaced by continuous variables and Gaussian mutations, which enable the $(1+1)$-ES to generate larger mutations and to escape from local optima [136].

The $(1+1)$-ES

The first ES, the so-called $(1+1)$-ES or two-membered evolution strategy, uses one parent and one offspring only. Two rules have been applied to these candidate solutions:

1. Apply small, random changes to all variables.
2. If the offspring solution is better (has a better function value) than the parent, take it as the new parent, otherwise retain the parent.

Schwefel [137] describes this algorithm as "the minimal concept for an imitation of organic evolution". The $(1 + 1)$-ES (Algorithm 12.1) is applied by many optimization practitioners to their optimization problem and included in our analysis for three reasons:

1. It is easy to implement,
2. it requires only a few exogenous parameters, and
3. it defines a standard for comparisons.

The symbol f denotes an objective function $f : \mathbb{R}^d \to \mathbb{R}$ to be minimized. The standard deviation σ will be referred to as *stepwidth* or *mutation strength*.

Algorithm 12.1 The two-membered evolution strategy or $(1 + 1)$-ES.

1: $t = 1$. Determine a point $X_1^{(t)}$ with associated position vector $x_1^{(t)} \in \mathbb{R}^d$, a standard deviation $\sigma^{(t)}$, and the function value $y_1 = f(x_1^{(t)})$.

$\qquad\qquad\qquad\qquad\qquad\qquad\qquad\qquad$ /* Initialization */

2: **while** some stopping criterion is not fulfilled **do**

3: $\quad$ Generate a new point $X_2^{(t)}$ with associated position vector $x_2^{(t)}$ as follows:

$$x_2^{(t)} = x_1^{(t)} + z \qquad\qquad (12.1)$$

$\quad$ where z is a d-dimensional vector. $\qquad\qquad\qquad$ /* Mutation */

4: $\quad$ Each component of z is the realization of a normal random variable Z with mean zero and standard deviation $\sigma^{(t)}$. $\qquad\qquad$ /* Evaluation */

5: $\quad$ Determine the function value $y_2 = f(x_2^{(t)})$.

6: $\quad$ Accept $X_2^{(t)}$ as $X_1^{(t+1)}$ if

$$y_2 < y_1 \qquad\qquad (12.2)$$

$\quad$ otherwise retain $X_1^{(t)}$ as $X_1^{(t+1)}$. $\qquad\qquad$ /* Selection */

7: $\quad$ Increment t.

$\qquad\qquad\qquad$ Update $\quad \sigma^{(t)} \qquad\qquad\qquad\qquad (12.3)$

$\qquad\qquad\qquad\qquad\qquad\qquad\qquad\qquad$ /* Adaptation */

8: **end while**

The convergence rate φ is defined as the expected distance traveled in the useful direction per mutation (iteration, generation), *i.e.*

$$\varphi = \int p(s)s\,ds \qquad\qquad (12.4)$$

where $p(s)$ is the probability for a mutation to cover a distance s towards the optimizer x^*. Rechenberg [136] presented an approximate analysis of the

$(1 + 1)$-ES and derived convergence-rate results for the corridor and sphere model. The corridor model defines a region of width b,

$$f_1(x) = c_0 + c_1 x_1 \tag{12.5}$$

with $|x_i| \leq b/2 \quad \forall i \in \{2, \ldots, d\}$. The sphere model is defined as

$$f_2(x) = c_0 + c_1 \sum_{i=1}^{d} (x_i - x_i^*)^2 \tag{12.6}$$

where x^* denotes the minimizer. Rechenberg's analysis showed, that

1. The optimal standard deviation σ_{opt} and maximum convergence rates are inversely proportional to the problem dimension d.
2. If the standard deviation of each component of the normally distributed vector z is continuously adjusted to the proper order of magnitude, then linear convergence order can be achieved.
3. The optimal mutation rate σ_{opt} corresponds to a success probability P_{succ} that is independent of d. The optimal success probability is approximatively $1/5$ for both models f_1 and f_2.

The following example illustrates Rechenberg's results.

Example 12.1 (Step-size adaptation). Based on the maximum progress rate

$$\varphi_{\max} = k_1 r / d, \qquad k_1 \simeq 0.2025 \tag{12.7}$$

with a common variance σ^2, which is always optimal given by

$$\sigma_{\mathrm{opt}} = k_2 r / d, \qquad k_2 \simeq 1.224 \tag{12.8}$$

for all components z_i of the random vector z (Equation (12.1)), the following relation can be derived:

$$\lim_{d \to \infty} \frac{\sigma_{\mathrm{opt}}^{(t+d)}}{\sigma_{\mathrm{opt}}^{(t)}} = \lim_{d \to \infty} \left(1 - \frac{k_1}{d}\right)^d = \exp^{-k_1} \simeq 0.817 \simeq \frac{1}{1.224} \tag{12.9}$$

Figure 12.1 visualizes the step-size adaptation process. The value from Equation (12.9) is used to adjust the standard deviation of the components of the d-dimensional random vector z. □

The $(1+1)$-ES requires the specification of five exogenous strategy parameters, or, statistically speaking, five factors, before the optimization run is started. The factors of the $(1 + 1)$-ES are summarized in Table 12.1.

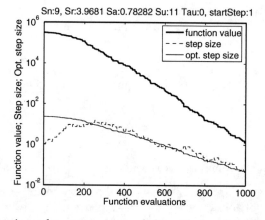

Fig. 12.1. Function values, step size, and theoretical optimal step size to demonstrate the self-adaptation process of the $(1+1)$-ES. The $(1+1)$-ES uses after 200 function evaluations the optimal step size (standard deviation), the adaptation process is successful.

Table 12.1. Factors of the two-membered evolution strategy. Based on the default values, the step size σ is multiplied by 0.85, if the success rate is larger than $1/s_r = 1/5$ or equivalently, if more than 20 out of 100 mutations have been successful.

Symbol	Factor	Range	Default
s_n	Adaptation interval	$\mathbb{N}$	100
s_r	1/success rate	$\mathbb{R}_+$	5
s_a	Step-size adjustment factor	$\mathbb{R}_+$	0.85
$\sigma^{(0)}$	Starting value of the step size σ	$\mathbb{R}_+$	1
$s_{1/5}$	Step-size update rule	$\{\text{intv, cont}\}$	cont

Multimembered Evolution Strategies

Rechenberg introduced the first multimembered ES, the so-called $(\mu+1)$-ES. It uses μ parents and one offspring and is referred to as the steady-state ES. Schwefel introduced the $(\mu+\lambda)$-ES, in which $\lambda \geq 1$ candidate solutions are created in a generation, and the best μ out of all $\mu+\lambda$ individuals survive, and the (μ, λ)-ES, in which the parents are forgotten and only the μ best out of λ candidate solutions survive. A birth surplus is necessary for the (μ, λ)-ES, that is, $\lambda > \mu$.

The multimembered evolution strategy can be described briefly as follows: The parental population is *initialized* at time (generation) $g = 0$. Then λ offspring individuals are generated in the following manner: A parent family of size ρ is selected randomly from the parent population. *Recombination* is applied to the object variables and the strategy parameters. The mutation operator is applied to the resulting offspring vector. After evaluation, a *selection* procedure is performed to determine the next parent population. The populations created in the iterations of the algorithm are called *generations* or *reproduction cycles*. A termination criterion is tested. If this criterion is not fulfilled, the generation counter (g) is incremented and the process continues with the generation of the next offspring. Figure 12.2 illustrates the basic EA loop. Because there are some fundamental differences between selection, recombination, and mutation in genetic algorithms and evolution strategies, these procedures will be detailed on the following. References [138] and [139] provide a comprehensive introduction to evolution strategies.

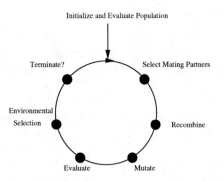

Fig. 12.2. EA loop. Evolution strategies use two kinds of selection procedures: (i) Mating selection determines parents randomly—in contrast to GA, which use fitness-based selection schemes—to generate offspring, and (ii) environmental selection, that can be interpreted as (fitness based) truncation selection.

Selection in Multimembered ES

Selection should direct the evolutionary search to promising regions. In ES, only candidate solutions with good function values are allowed to reproduce. The mating selection process is deterministic in contrast to the random processes used in genetic algorithms (cf. Section 11.2). This selection scheme is known as *truncation* or *breeding selection* in biology.

The κ-selection scheme takes the age of the candidate solutions into account: only candidate solutions that are younger than κ are allowed to reproduce.

For $\kappa = 1$ this selection method is referred to as *comma-selection*: only offspring individuals can reproduce. Consider the following situations:

- If $\mu = \lambda$, then all offspring are selected and the population performs a random walk, because the selection process provides no information to guide the search.
- A necessary condition of convergence towards an optimal solution is $\mu < \lambda$.

The κ-selection is referred to as *plus-selection* for $\kappa = \infty$: both the offspring and the parents belong to the mating pool. The plus-selection is an elitist selection scheme, because it guarantees the survival of the best individual found so far. *Steady-state*-ES that use μ parents and one offspring are used in asynchronous parallel systems.

Variation in Multimembered ES

Reference [139] proposes some guidelines derived from the philosophy of Darwinian evolution to design variation operators.

1. A *state* comprises a set of object and strategy parameter values ($x^{(t)}$, $s^{(t)}$). *Reachability* demands that any state can be reached within a finite number of iterations. This feature is necessary to prove (theoretically) global convergence.
2. Variation operators (mutation and recombination) should not introduce any bias, *e.g.* by considering only good candidate solutions. Variation operators are designed to *explore* the search space in contrast to selection operators that exploit the gathered information. Table 12.2 summarizes the features of the ES operators.
3. *Scalability* is the third criterion that should be fulfilled by variation operators in ES: Small changes of the representation should cause small changes in the function values.

Table 12.2. Guidelines from the philosophy of Darwinian evolution in ES

Operator	Creation	Information
Variation	Stochastically	Unbiased
Selection	Deterministically	Biased (use function values)

The standard ES recombination operators produce from a family of ρ parent individuals one offspring. Note that the standard GA crossover operator uses two parents to create two offspring.

Consider a set of parental vectors representing either object or strategy parameters:

$$\{(x_{11}, \ldots, x_{1d}), (x_{21}, \ldots, x_{2d}), \ldots, (x_{\mu 1}, \ldots, x_{\mu d})\} \qquad (12.10)$$

Two recombination schemes are commonly used in ES. Both use a set $\mathcal{R} = \{r_1, r_2, \ldots, r_\rho\}$, that represents the indices of the mating partners. It is constructed by randomly (uniformly) choosing ρ values (without replacement) from the set $\{1, 2, \ldots, d\}$.

1. *Discrete recombination* can be described as follows. The recombined vector is determined as

$$x_{\text{discrete}} = (x_{u_11}, x_{u_22}, \ldots, x_{u_dd}\} \tag{12.11}$$

where u_i is randomly (uniformly) chosen from $\mathcal{R}$.

2. To implement *intermediate recombination*, Equation (12.11) has to be modified as follows:

$$x_{\text{intermediate}} = \frac{1}{\rho} \left(\sum_{i=1}^{\rho} x_{r_i1}, \sum_{i=1}^{\rho} x_{r_i2} \ldots, \sum_{i=1}^{\rho} x_{r_id} \right) \tag{12.12}$$

Mutation—which can be considered as the basic variation operator in evolution strategies—is applied to the recombined solution, cf. Fig. 12.2. Mutation in ES is a self-adaptive process that relies on the individual coupling of endogenous strategy parameters with object parameters. After being recombined as described above, the strategy parameters (standard deviations, step widths, or mutation strengths) are adapted. These adapted values are used in a second step to mutate the object parameters.

We consider algorithms with one stepwidth σ first. To prevent negative standard deviations, the adaptation of the step width must be multiplicative. Reference [139] discusses an additional argument for a multiplicative mutation of the mutation strength on the sphere model f_2, see Equation (12.6). It can be shown, that in expectation σ should be changed by a constant factor. Therefore, the mutation operator can be implemented as

$$\sigma^{(t+1)} = \sigma^{(t)} \cdot \exp(\tau z) \tag{12.13}$$

where z is a realization of the $\mathcal{N}(0,1)$ distributed random variable Z. The parameter τ is the so-called *learning rate*. The object variables are mutated next:

$$x^{(t+1)} = x^{(t)} + w \tag{12.14}$$

where w is a realization of the $\mathcal{N}(0, \sigma^{(t+1)})$ distributed random variable W. Taking the logarithm on both sides of Equation 12.13 results in

$$\log(\sigma^{(t+1)}) = \log(\sigma^{(t)}) + \tau z \tag{12.15}$$

Hence, mutations of the strategy (on a logarithmic scale) and object parameters are structurally similar. This multiplicative mutation scheme for one σ can be extended to several strategy parameters $\sigma = (\sigma_1, \ldots, \sigma_d)$. Reference [140] proposes the following extended log-normal rule:

$$\sigma^{(t+1)} = \exp(\tau_0 z_0) \cdot \left(\sigma_1^{(t)} \exp(\tau z_1), \ldots, \sigma_d^{(t)} \exp(\tau z_d) \right) \tag{12.16}$$

where z_i are realizations of the $\mathcal{N}(0,1)$ distributed random variables Z_i, $i = 1, \ldots, d$. This mutation scheme uses a global and a local learning parameter τ_0 and τ respectively. The whole vector is scaled by the random factor $\exp(\tau_0 z_0)$ after each component has been mutated.

12.1.2 Algorithm Designs for Evolutionary Algorithms

We consider the parameters or control variables from Table 12.3. This table shows typical parameter settings. The reader is referred to Ref. [141] for a detailed description of these parameters.

Table 12.3. Default settings of exogenous parameters of a "standard" evolution strategy. Problems may occur, when these "standards" are not adjusted to the specific optimization problem.

Symbol	Parameter	Range	Default
μ	Number of parent individuals	$\mathbb{N}$	15
$\nu = \lambda/\mu$	Offspring–parent ratio	$\mathbb{R}_+$	7
$\sigma_i^{(0)}$	Initial standard deviations	$\mathbb{R}_+$	3
n_σ	Number of standard deviations. d denotes the problem dimension	$\{1, d\}$	1
c_τ	Multiplier for individual and global mutation parameters	$\mathbb{R}_+$	1
ρ	Mixing number	$\{1, \mu\}$	2
r_x	Recombination operator for object variables	$\{i, d\}$	d (discrete)
r_σ	Recombination operator for strategy variables	$\{i, d\}$	i (intermediary)
κ	Maximum age	$\mathbb{R}_+$	1

After this short introduction to ES, we consider noisy optimization problems. Related problems for GA have been discussed in Section 11.3.

12.2 Optimization of Noisy Fitness with Evolution Strategies

In many real-world optimization problems, function values can only be estimated but not determined exactly. Falsely calibrated measurement instruments, inexact scales, scale reading errors, *etc.* are typical sources for measurement errors. If the function of interest is the output from stochastic simulations, *e.g.* transportation systems, then the measurements may be exact, but some of the model output variables are random variables.

Computer simulations are a suitable means to optimize many actual real-world problems (*optimization via simulation*). We will concentrate on stochastic simulation models in the following. The reader is referred to Ref. [142] for deterministic models.

For example, consider elevator movements in high-rise buildings. As the result of a simulation run is a random variable, we cannot optimize the actual value of the simulation output, or a singular performance of the system Y. One goal of optimization via simulation is to optimize the expected performance $E[Y(x_1, x_2, \ldots, x_n)]$, where the x_is denote the controllable input variables [143–145].

The stochastic output in optimization via simulation complicates the selection process in direct search methods. The efficiency of the evaluation and selection method is a crucial point, since the search algorithm may not be able to make much progress if the selection procedure requires many function evaluations. From our point of view the following case is fundamental for the selection procedure in noisy environments:

> Reject or accept a new candidate, while the available information is uncertain. Thus, two errors may occur: An α error as the probability of accepting a worse candidate due to noise and a β error, as the error probability of rejecting a better candidate.

The terms "candidate" and "point" will be used synonymously. A well-established context where these error probabilities are analyzed is hypothesis testing. We concentrate our investigations on the selection process when the function values are disturbed by additive noise, cf. Equation (11.9). Noise that affects the object variables is not the subject of our investigations. Our analysis is based on the following statistical assumptions. Let $\{Y_{ij}\}$, $1 \leq i \leq r$, $1 \leq j \leq k$, denote r independent random samples of observations, taken from $k \geq 2$ candidates. The Y_{ij} can denote function values taken from candidate solutions $X_1, \ldots, X_k$ or individuals (particles) of some evolutionary algorithm. Candidate X_i has a (fitness) function value with unknown mean μ_i and common unknown variance $\sigma_{\epsilon,i}^2 = \sigma_\epsilon^2$, $1 \leq i \leq k$. The *ordered means* are denoted by

$$\mu_{[1]} \leq \mu_{[2]} \leq \cdots \leq \mu_{[k]} \qquad (12.17)$$

where $\mu_{[1]}$ denotes the mean of the best candidate (minimization). Generally, normal response experiments are considered.

12.2.1 Ways to Cope with Uncertainty

Noise makes it difficult to compare different solutions and select the better ones. Noise affects the selection process in evolutionary algorithms: In every iteration, the μ best out of λ candidate solutions have to be determined.

Wrong decisions can cause a *stagnation* of the search process: Overvaluated candidates—solutions that are only seemingly better—build a barrier around the optimum and prevent convergence. The function value at this barrier will be referred to as the *stagnation level*. Or, even worse, the search process can be *misguided*: The selection of seemingly good candidates moves the search away from the optimum. This phenomenon occurs if the noise level is high and the probability of a correct selection is very small.

Common Techniques

One may attempt to reduce the effect of noise explicitly. The simplest way to do so is to sample a solution's function value n times, and use the average as an estimate for the true expected function value. This reduces the standard deviation of the noise by a factor of $\sqrt{n}$, while increasing the running time by a factor of n.

Further means used by evolutionary algorithms to cope with noise are averaging techniques based on statistical tests, local regression methods for function value estimation, or methods to vary the population size [105, 146–150]. It has been shown for evolution strategies that increasing the population size may help the algorithm to cope with the noise [151]. A comprehensive overview of various evolutionary algorithm variants in the presence of noise is given in [152].

A Taxonomy of Selection Methods

As introduced above, noise affects the selection process of evolutionary algorithms. In the following, a comprehensive taxonomy of elementary selection methods is developed. Depending on the prior knowledge, selection schemes can be classified according to the following criteria:

Threshold: subset selection – indifference zone.
Termination: single stage – multi stage (sequential).
Sample size: open procedures – closed procedures.
Variances: known – unknown, equal – unequal.

Subset selection is used when analyzing results, whereas the *indifference zone* (IZ) approach is used when designing experiments. Subset selection and IZ approaches use a threshold value. The goal of subset selection is the identification of a subset containing the best candidate. It is related to screening procedures. The sample size r is known in subset selection approaches, it is determined prior to the experiments in the indifference zone approaches.

Single stage procedures can be distinguished from *multi stage* procedures. The terms "multi stage" and "sequential" will be used synonymously. Single-stage procedures can use *elimination*: If inferior solutions are detected, they are eliminated immediately. Selection procedures are *closed*, if prior to experimentation an upper bound is placed on the number of observations to be taken from each candidate. Otherwise, they are *open*. Furthermore, it is important to know whether the variance is common or known.

Sophisticated Selection Methods

More sophisticated sampling approaches include the information about variances and the desired probability of a correct selection and adapt the number of samples according to this information. Two-stage procedures use the first stage to estimate variances. In the second stage an additional number of samples is drawn for each candidate solution, each number depending on the variance and the overall required probability of correct selection. *Sequential procedures* allow even more than two stages. We consider procedures that assign a fixed total number of samples to candidate solutions, but sequentially decide how to allocate the samples on the different candidate solutions in Section 12.2.2. Such methods use either an elimination mechanism to reduce the number of alternatives considered for sampling, or they assign additional samples only to the most promising alternatives.

We will present two modern approaches for direct search algorithms to cope with noise: The optimal computing budget allocation is presented in Section 12.2.2, threshold selection is introduced in Section 12.2.3.

12.2.2 Optimal Computing Budget Allocation

A recently suggested sequential approach is the *optimal computing budget allocation* (OCBA) [153, 154]. The aim of this method is to find the best within a set of candidate solutions and to select it with a high probability. The OCBA variant is a closed procedure and can be used for unknown and unequal variances. It draws samples sequentially until the computational budget is exhausted while adjusting the selection of samples to maximize the probability of a correct selection. The event of selecting the best is denoted by CS (correct selection), and the probability of this event $P(CS)$. The following notation is used:

- X_{ij}: the observation from the jth replication of the ith design,
- N_i: the final number of replications for design i,
- r: the intermediate number for design i,
- μ_i: the expected performance measure for design i, *i.e.* $\mu_i = E(X_{ij})$,
- $\bar{X}_i$: the running sample mean for design i, *i.e.* $\sum_{j=1}^{r} \frac{X_{ij}}{r}$,
- X_i: the final sample mean for design i, *i.e.* $\sum_{j=1}^{N_i} \frac{X_{ij}}{N_i}$,
- σ_i^2: the variance of the observed performance measure of design i from one replication, *i.e.* $\sigma_i^2 = Var(X_{ij})$
- $S_i^2(N_i)$: the sample variance of design i with N_i replications, *i.e.* $S_i^2(N_i) = \sum_{j=1}^{N_i} \frac{(X_{ij} - \bar{X}_i)^2}{N_i - 1}$

Based on a Bayesian model, Chen *et al.* [153] develop an *approximate probability of correct selection* (APCS). They show that the APCS is a lower bound of the $P(CS)$.

$$P(CS) \geq \underbrace{1 - \sum_{i=1, i \neq b}^{k} P[\bar{X}_b > \bar{X}_i]}_{APCS},$$

where b denotes the actually best design. For a fixed number of replications or batches, the APCS can be asymptotically maximized when

$$\frac{N_i}{N_j} = \frac{\sigma_i/(\bar{X}_i - \bar{X}_b)}{\sigma_j/(\bar{X}_j - \bar{X}_b)}, i, j \in 1, 2, \ldots, k, \text{ and } i \neq j \neq b \qquad (12.18)$$

$$N_b = \sigma_b \sqrt{\sum_{i=1, i \neq b}^{k} \frac{N_i^2}{\sigma_i^2}} \tag{12.19}$$

where σ_i is the standard deviation of the response of design i.

The OCBA procedure as described in Algorithm 12.2 uses the following parameters: T is the total budget and n_0 the initial number of samples. It adjusts the selection of samples to approximate Equation (12.18) and Equation (12.19). For a more detailed description we refer the reader to Ref. [153].

Algorithm 12.2 OCBA Procedure.

1: Simulate n_0 initial replications for each design.
Set $l = 0, N_1^l = N_2^l = \cdots = N_k^l = n_0$, and $T = T - kn_0$ /* Initialization */
2: Set $l = l + 1$. Increase the computing budget by Δ_l (i.e. number of additional simulations) and compute the new budget allocation to approximate Equation (12.18) and Equation (12.19).   /* Budget determination */
3: Simulate additional $\max(0, N_i^l - N_i^{l-1})$ replications for each design $i, i = 1, 2, \ldots, k$ /* Replication */
4: $T = T - \Delta_l$. If $T > 0$, go to step 2   /* Update total budget */
5: Return the index b of the system with the lowest mean $\bar{X}_b$, where $\bar{X}_b = \min_{1 \leq i \leq k} \bar{X}_i$ /* Result */

OCBA can be integrated in the selection procedure of ES and PSO. In [154] it was demonstrated that OCBA performs well on simple test problems. Further investigations on complex real-world optimization problems will be analyzed in Chapter 16.

12.2.3 Threshold Selection

Threshold rejection (TR) and *threshold acceptance* (TA) are complementary strategies. Threshold rejection is a selection method for evolutionary algorithms that accepts new candidates if their noisy function values are significantly better than the value of the other candidates [155]. "Significant" is equivalent to "by at least a margin of τ". Threshold acceptance accepts a new candidate even if its noisy function value is worse. The term *threshold selection* (TS) subsumes both selection strategies. TS should not be confused with "tabu search". The basic idea of TS is also known in other contexts like plant and animal breeding: "Only individuals with a phenotypic value at least as great as some number c are permitted to reproduce." Truncation selection is important for breeders, but it is unlikely to occur in natural populations. Threshold selection is also related to Fredkin's paradox: "The more equally attractive two alternatives seem, the harder it can be to choose between them—no matter that, to the same degree, the choice can only matter less" [156].

The Threshold Selection Procedure

The experimental goal is to select the candidate associated with the smallest mean $\mu_{[1]}$. This basic procedure as shown in Algorithm 12.3 can be implemented in many optimization algorithms, for example evolution strategies or particle swarm optimization. As can be seen from Equation (12.20), threshold

---

Algorithm 12.3 Threshold selection.

1: Given: A candidate X_1 with a related sample Y_{1j} of r observations and sample mean $\overline{y}_1 = \sum_{j=1}^{r} y_{1j}/r$.
2: Take a random sample of r observations Y_{2j}, $1 \le j \le r$, in a single stage from a new candidate X_2.
3: Calculate the sample mean $\overline{y}_2 = \sum_{j=1}^{r} y_{2j}/r$.
4: Select the new candidate X_2 if and only if

$$TR : \overline{y}_2 + \tau < \overline{y}_1, \text{ with } \tau \ge 0 \tag{12.20}$$

or

$$TA : \overline{y}_2 + \tau < \overline{y}_1, \text{ with } \tau \le 0 \tag{12.21}$$

---

rejection increases the chance of rejecting a worse candidate at the expense of accepting a good candidate. It might be adequate if there is a very small probability of generating a good candidate. Equation (12.21) reveals that threshold acceptance increases the chance of accepting a good candidate at the risk of failing to reject worse candidates.

Threshold Selection and Hypothesis Testing

The determination of a threshold value can be interpreted in the context of classical hypothesis testing as the determination of a critical point [157]. The critical point $c_{1-\alpha}$ for a hypothesis test is a threshold to which one compares the value of the test statistic in a sample. It specifies the *critical region* CR and can be used to determine whether or not the null hypothesis is rejected. We are seeking a value $c_{1-\alpha}$, such that

$$Pr\{S > c_{1-\alpha} \quad |H \text{ true }\} \le \alpha \tag{12.22}$$

where S denotes the test statistic and the null hypothesis H reads: "There is no difference in means." The threshold acceptance selection method can be interpreted in a similar manner.

Known Theoretical Results

The theoretical analysis in [155], where threshold rejection was introduced for evolutionary algorithms with noisy function values, was based on the progress rate theory on the sphere model and was shown for the $(1 + 1)$-evolution strategy. However, this theoretical result is only applicable when the distance to the optimum and the noise level are known. These are conditions that are not very often met in practice.

By interpreting this result qualitatively, we can see that the threshold value τ should be increased while approaching the optimizer x^* (τ should be infinite, when the optimum is obtained). Another approach will be presented in the following example. Here, we will demonstrate how threshold rejection can improve the quality gain [157]. The quality gain is defined as the expected change in the function value. The influence of TR on the selection process was analyzed using a simple stochastic search model that is related to a model proposed in [83]. Then the search can be misled, although the algorithm selects only "better" candidates. TR can prevent this effect. In the simple stochastic search model the optimal threshold value can be calculated as a function of the noise strength, the probability of generating a better candidate, and the difference between the expectation of the function values of two adjacent states.

Example 12.2 (A Simple Model of Stochastic Search). We analyze the influence of TS on the selection process in a simple stochastic search model. This model possesses many crucial features of real-world optimization problems, *i.e.* a small probability of generating a better offspring in an uncertain environment. The following simple stochastic search model is considered:

Suppose that the system to be optimized is at time t in one of the consecutive discrete states $X_t = i$, $i \in \mathbb{Z}$. In state i, we can probe the system to obtain a fitness value $\tilde{f}(X_t) = i \cdot \delta + U$, where $\delta \in \mathbb{R}^+$ represents the distance between the expectation of the fitness values of two adjacent states. The random variable U possesses normal $\mathcal{N}(0, \sigma_\epsilon^2)$ distribution. The goal is to take the system to a final state $X_t = i$ with i as high as possible (maximization problem) in a given number of steps.

Next, we consider a simple search with TS for this search model, see Algorithm 12.4. In this simple search algorithm, the probability of generating a better candidate p is given. In general, the experimenter has no control over p, which would be some small value for nontrivial optimization tasks.

This simple search algorithms can be represented by a Markov chain $\{X_t\}$ with the following properties:

1. $X_0 = 0$.
2. $P\{X_{t+1} = i + 1 | X_t = i\} = p \cdot P_\tau^+$
3. $P\{X_{t+1} = i - 1 | X_t = i\} = (1 - p) \cdot (1 - P_\tau^-)$
4. $P\{X_{t+1} = i | X_t = i\} = p \cdot (1 - P_\tau^+) + (1 - p) \cdot P_\tau^-$ with

$$P_\tau^\pm := \Phi\left(\frac{\delta \mp \tau}{\sqrt{\frac{m+n}{mn}} \sigma_\epsilon}\right). \tag{12.24}$$

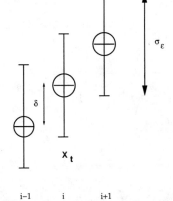

Fig. 12.3. Simple stochastic search. Adjacent states.

Algorithm 12.4 Simple search algorithm. An implementation of this search
model in C is shown in the chapter 'Program Listings' at the end of this book

1: Initial state $X_{t=0} = 0$.   /* *Initialize* */
2: At the tth step, with current state $X_t = i$, flip a biased coin: Set the candidate
   of the new state Y_t to $i + 1$ with probability p and to $i - 1$ with probability
   $(1 - p)$.   /* *Generate offspring* */
3: Draw samples (fitness values) from the current and the candidate states:

$$\tilde{f}(X_{t,j}) \text{ and } \tilde{f}(Y_{t,k}) \qquad (12.23)$$

   with the measured fitness value $\tilde{f}(X) := f(X) + w$, where w is the realization
   of a random variable, representing normal distributed noise, $W \sim \mathcal{N}(0, \sigma_\epsilon^2)$. /*
   Evaluate */
4: Determine a threshold value τ. If $\overline{f}(Y_t) + \tau > \overline{f}(X_t)$, accept Y_t as the next state:
   $X_{t+1} := Y_t$; otherwise , keep the current state: $X_{t+1} := X_t$.   /* *Select* */
5: If $t < t_{\max}$, increment t and go to step 2.   /* *Terminate* */

The measurement of the local behavior of an EA can be based on the expected
distance change in the object parameter space from generation to generation.
This leads to the following definition of the quality gain: Let R be the number
of advance in the state number t in one step:

$$R := X_{t+1} - X_t \qquad (12.25)$$

The *quality gain* is defined as the expectation

$$E[R(\delta, \sigma_\epsilon, p, t)] \qquad (12.26)$$

to be abbreviated $E[R]$. Based on the Markov model (Equation (12.24)) it
can be shown that

$$E[R_\tau] = p \cdot P_\tau^+ - (1 - p) \cdot (1 - P_\tau^-) \qquad (12.27)$$

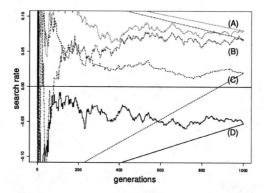

Fig. 12.4. Simple stochastic search. Simulations performed to analyze the influence of TS on the search rate. (C) and (D) show results for an optimal threshold value and a zero threshold value, respectively. (A) and (B) represent the corresponding graphs for increased success probabilities(from $p = 0.3$ to 0.4).

In this simple stochastic search model it is possible to determine the optimal τ_{opt} value with regard to the quality gain, if the fitness function is disturbed with normal-distributed noise:

$$\tau_{opt} = \frac{\sigma_\epsilon^2}{\delta} \log \frac{1-p}{p} \qquad (12.28)$$

Assume there is a very small success probability p. Then the search can be misled, although the algorithm selects only "better" candidates. We can conclude from Equation (12.27), that a decreasing success probability ($p \searrow 0$) leads to a negative quality gain. Based on Equation (12.28), we calculated the optimal threshold value for 5 different success probabilities to illustrate the influence of TS on the quality gain, cf. Table 12.4. Corresponding values of the quality gain are shown in the third column. TS can enhance the quality gain and even avoid a negative quality gain. This can be seen from the values in the last column.

Figure 12.4 reveals that simulations lead to the same results. For two different p-values, the influence of TS on the quality gain is shown. The quality

Table 12.4. Simple stochastic search. The noise level σ_ϵ equals 1.0, the distance δ is 0.5. Labels (A) to (D) refer to the results of the corresponding simulations shown in Fig. 12.4.

p	τ_{opt}	$E[R_{\tau=0}]$	$E[R_{\tau_{opt}}]$
0.1	4.394	-0.262	0.00005
0.2	2.773	-0.162	0.003
0.3	1.695	-0.062 (D)	0.018 (C)
0.4	0.811	0.038 (B)	0.059 (A)
0.5	0.0	0.138	0.138

gain becomes negative, if p is set to 0.3 and no TS is used (D). The situation can be improved, if we introduce TS: The quality gain becomes positive (C). A comparison of (B), where a zero threshold was used, and (A), where the optimal threshold value was used, shows that TS can improve an already positive quality gain. These results are in correspondence with the theoretical results in Table 12.4.   □

Another interesting aspect of threshold selection is related to the curvature of the objective function: Recall that a convex function is a continuous function whose value at the midpoint of every interval in its domain does not exceed the average of its values at the ends of the interval. Thus, $f(x)$ is *convex* on an interval $[a, b]$ if for any two points x_1 and x_2 in $[a, b]$,

$$f\left(\frac{1}{2}(x_1 + x_2)\right) \leq f(x_1 + x_2)$$

A function $f(x)$ is *strictly convex* if

$$f\left(\frac{1}{2}(x_1 + x_2)\right) < f(x_1 + x_2)$$

Let f denote a strictly convex test function. Threshold acceptance produces better quality results than plus selection in the vicinity of a local minimum of f if the function values are disturbed by additive, Gaussian noise. The optimal threshold value increases as the noise level grows. A similar result is true for strictly concave functions and threshold rejection in the vicinity of a local maximum. Further details of the relationship between curvature and threshold selection are discussed in [158].

12.3 Particle Swarm Optimization

12.3.1 Framework of Particle Swarm Optimization

The main inspiration, which led to the development of *particle swarm optimization* (PSO) algorithms, was the flocking behavior of swarms and fish shoals [159]. PSO has been applied to numerous simulation and optimization problems in science and engineering [160–162].

PSO belongs to the class of stochastic, population-based optimization algorithms [160]. It exploits a population of individuals to probe the search space. In this context, the population is called a *swarm* and the individuals are called *particles*. Each particle moves with an adaptable velocity within the search space, and it retains in a memory the best position it has ever visited.

A PSO's convergence is controlled by a set of design variables that are usually either determined empirically or set equal to widely used default values.

There are two main variants of PSO with respect to the information exchange scheme among the particles. In the *global* variant, the best position ever attained by all individuals of the swarm is communicated to all the particles at each iteration. In the *local* variant, each particle is assigned to a neighborhood consisting of prespecified particles. In this case, the best position ever attained by the particles that comprise a neighborhood is communicated among them. Neighboring particles are determined based on their indices rather than their actual distance in the search space. In the current work we consider the global PSO variant only.

Assume a d-dimensional search space, $S \subset \mathbb{R}^d$, and a swarm consisting of s particles. The ith particle is a d-dimensional vector,

$$x_i = (x_{i1}, x_{i2}, \ldots, x_{id})^T \in S$$

The velocity of this particle is also a d-dimensional vector,

$$v_i = (v_{i1}, v_{i2}, \ldots, v_{id})^T$$

The best previous position encountered by the ith particle (*i.e.* its memory) in S is denoted by

$$p_i^* = (p_{i1}^*, p_{i2}^*, \ldots, p_{id}^*)^T \in S$$

Assume b to be the index of the particle that attained the best previous position among all the particles in the swarm, and t to be the iteration counter.

Particle Swarm Optimization with Inertia Weights

The resulting equations for the manipulation of the swarm in the PSO with inertia weights are [163],

$$v_i(t+1) = wv_i(t) + c_1 r_1 \left(p_i^*(t) - x_i(t)\right) + c_2 r_2 \left(p_b^*(t) - x_i(t)\right) \quad (12.29)$$
$$x_i(t+1) = x_i(t) + v_i(t+1) \quad (12.30)$$

where $i = 1, 2, \ldots, s$; w is a parameter called the *inertia weight*; c_1 and c_2 are positive constants, called the *cognitive* and *social* parameter, respectively; and r_1, r_2 are vectors with components uniformly distributed in $[0, 1]$. All vector operations are performed componentwise.

Usually, the components of x_i and v_i are bounded as follows,

$$x_{\min} \leqslant x_{ij} \leqslant x_{\max}, \quad -v_{\max} \leqslant v_{ij} \leqslant v_{\max}, \quad j = 1, \ldots, n \quad (12.31)$$

where $x_{\min}$ and $x_{\max}$ define the bounds of the search space, and $v_{\max}$ is a parameter that was introduced in early PSO versions to avoid swarm explosion that was caused by the lack of a mechanism for controlling the velocity's magnitude. Although the inertia weight is such a mechanism, empirical results have shown that using $v_{\max}$ can further enhance the algorithm's performance.

Table 12.5. Default algorithm design $x_{\mathrm{PSO}}^{(0)}$ of the PSO algorithm. Similar designs have been used in [164] to optimize well-known benchmark functions.

Symbol	Parameter	Range	Default	Constriction
s	Swarm size	$\mathbb{N}$	40	40
c_1	Cognitive parameter	$\mathbb{R}_+$	2	1.494
c_2	Social parameter	$\mathbb{R}_+$	2	1.494
$w_{\max}$	Starting value of the inertia weight w	$\mathbb{R}_+$	0.9	0.729
w_{scale}	Final value of w in percentage of $w_{\max}$	$\mathbb{R}_+$	0.4	1.0
$w_{\mathrm{iterScale}}$	Percentage of iterations, for which $w_{\max}$ is reduced	$\mathbb{R}_+$	1.0	0.0
$v_{\max}$	Maximum value of the step size (velocity)	$\mathbb{R}_+$	100	100

Table 12.5 summarizes the design variables of particle swarm optimization algorithms.

Experimental results indicate that it is preferable to initialize the inertia weight with a large value, in order to promote global exploration of the search space, and gradually decrease it to get more refined solutions. Thus, an initial value around 1 and a gradual decline towards 0 is considered a proper choice for w. This scaling procedure requires the specification of the maximum number of iterations $t_{\max}$. Reference [165] illustrates a typical implementation of this scaling procedure.

Proper fine tuning of the parameters may result in faster convergence and alleviation of local minima [163, 165–167]. Different PSO versions, such as PSO with a constriction factor, have been proposed [168].

Particle Swarm Optimization with Constriction Coefficient

In the constriction factor variant, Equation (12.29) reads,

$$v_i(t+1) = \chi \left[v_i(t) + c_1 r_1 \left(p_i^*(t) - x_i(t) \right) + c_2 r_2 \left(p_b^*(t) - x_i(t) \right) \right] \qquad (12.32)$$

where χ is the *constriction factor* [169].

Equations (12.29) and (12.32) are similar. In our experiments, the so-called *canonical* PSO variant proposed in [169], which is the constriction variant defined by Equation (12.32) with $c_1 = c_2$, has been used. The corresponding parameter setting for the constriction factor variant of PSO is reported in the last column (denoted as "Constriction") of Table 12.5, where χ is reported in terms of its equivalent inertia weight notation, for uniformity reason. Reference [170] gives an overview of current PSO variants.

12.3.2 PSO and Noisy Optimization Problems

PSO in the presence of noise has been examined before by Parsopoulos and Vrahatis [171], and Krink *et al.* [172]. We extend these studies by also examining the influence of algorithm parameters, by considering a wider spectrum of

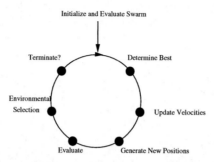

Fig. 12.5. The PSO loop is similar to the EA loop from Fig. 12.2. PSO determines the positions of the local and the global best particles. These values are used to update the velocities and to generate new positions. Note, that the update of the velocities in PSO as well as recombination and mutation in EA are stochastic procedures.

Table 12.6. Default settings of the exogenous parameters of PSO with constriction factor. Recommendations from Ref. [168].

Symbol	Parameter	Range	Default
s	Swarm size	$\mathbb{N}$	40
χ	Constriction coefficient	$\mathbb{R}_+$	0.729
φ	Multiplier for random numbers	$\mathbb{R}_+$	4.1
$v_{\max}$	Maximum value of the step size (velocity)	$\mathbb{R}_+$	100

noise levels and different types of noise (multiplicative and additive). Experimental results from PSO with OCBA selection will be presented and discussed in Chapter 16.

As PSO uses global and local information to determine the new candidate solutions, it is interesting to analyze which information is of greater importance under noise. Therefore we investigated "global versus local certainty." As described above, noise can affect PSO in two steps: when each particle chooses its local best, and when the global best is determined. In order to analyze which of the two steps is more critical for algorithm performance, we made a preliminary study with two special variants of the PSO algorithm. Variant PSO_{pc} was given the correct information whenever a particle decided between its new position and the old personal best. This variant was able to find significantly better results than the variant $PSO_{default}$. Variant PSO_{gc} was provided with the information which of the particles' presumed best was the true best, *i.e.* it could correctly select the global best from all the presumed local best. In our experiments with additive and multiplicative noise models, PSO_{pc} showed clearly superior performance compared to the PSO_{gc} variant. However, we have to keep in mind that variant PSO_{pc} received in each iteration the knowledge for a number of decisions, which was equal to the swarm size. In contrast variant PSO_{gc} could decide once per iteration correctly. Furthermore, PSO_{gc} could potentially lose the true global best, namely when the decision on the first level was wrong. This can not happen with PSO_{pc}.

12.4 Summary

ES and PSO are population-based search algorithms. Both make use of randomness to generate new candidate solutions. PSO uses a stochastically disturbed weighted sum of the local (personal) and the global best position in the search space, whereas ES use bioinspired methods: recombination and mutation. The selection procedure, *i.e.* the update of the local and the global best positions for PSO, and the truncation-selection procedure in ES, can be disturbed if the function values cannot be determined exactly. We presented a taxonomy of methods to cope with noise and discussed two methods:

1. threshold selection for ES, and
2. OCBA for PSO.

The determination of an optimal threshold value in the simple search model required information that is usually not directly available in real-word situations and can only be estimated. For example, the probability of generating a better offspring is unknown during the search process.

In addition to restrictive assumptions in the noise model (constant noise level, Gaussian noise) our analysis considered the local performance only. Local performance measures are essential because evolution strategies operate locally. The probability to generate an offspring that is better than its parent decreases if the mutation rate is increased. [173] introduced the concept of the *evolution window*: It states that evolutionary progress occurs only within a very narrow band of the mutation step size σ.

Although local performance measures may be useful to predict the global performance, *e.g.* for the sphere model, the situation in real-world scenarios may be different. The chance of finding an absolute optimum of f among several local minima depends on the structure of its fitness landscape.

The analytical approximation requires the identification of relevant factors as in the experimental analysis, consider for example the selection phase for regression trees or the forward selection mechanisms for regression models. Reference [151] notes that this selection process is the "source for a more in-depth understanding of the functioning of ES."

Regarding the influence of threshold selection on the global performance, we will use the statistical selection methods introduced in previous chapters of this book.

Chapter 16 is devoted to the question of how to implement new selection operators and to draw scientifically meaningful conclusions about their performances.

Intelligent Control by Combinatorial Optimization

Many problems in transportation can be formulated as scheduling problems. As indicated earlier in Chapter 8, scheduling problems are usually computationally intensive, and often NP-hard; thus it is not a trivial task finding efficient algorithms for their solution.

In practice it is often more useful to arrive at an *approximate solution* within reasonable time constraints than attempting to find exact optimal solutions at any cost. Furthermore, since developing specialized heuristics for each specific problem would take too much effort, it is important to find generic methods, capable of solving problem classes.

Thus scheduling research, and generally many other branches of the combinatorial optimization research, have developed *metaheuristic algorithms* for the efficient, but usually approximate, solution of classes of discrete optimization problems.

Evolutionary algorithms, which are currently perhaps the most popular metaheuristics, are discussed extensively in Chapters 11 and 12. Of the many other metaheuristic methods, we introduce here two that were found useful in solving transportation scheduling problems: *Branch-and-Bound Search* and *Tabu Search*.

13.1 Branch-and-Bound Search

Branch-and-bound search is a general technique for obtaining the exact optimal solution for combinatorial optimization problems. Let us consider an optimization problem

$$(P) \quad x^* = \arg \min_{x \in X} f(x)$$

We call a candidate solution a node, and conceptually we consider the nodes arranged in a search tree, *i.e.* in a connected, directed acyclic graph.

Branch-and-bound search consists of

Branching Operation: to divide the search space into several subspaces, and reduce the original optimization problem to a sequential search of the subproblems.

Bounding Operation: by relaxing the problem to an optimization problem rather easy to solve, obtain lower bounds of the optimal values. If the optimal value is worse than the tentative best solution (referred to as the *incumbent solution*), the (sub)problem is not worth solving, and hence it is not decomposed further, *i.e.* fathomed. If the optimal value is better than the incumbent best solution, and the solution of the relaxed problem belongs to the search space of the subproblem, it is the optimal solution of the subproblem and we can update the incumbent solution. Otherwise the branching and bounding operation is applied to the subproblem recursively.

Given a node set X, an objective function $f(x)$, and an incumbent solution $\tilde{x}$, the prototype algorithm of the branch-and-bound search $BAB(X, f(x), \tilde{x})$ is as follows:

Algorithm 13.1 The prototype branch-and-bound search algorithm

---

1: Obtain the relaxed problem

$$(P')\quad x^*_{\text{relax}} = \arg\min_{x \in Y} f(x)$$

where $Y \supset X$. Typically it is obtained by relaxing the constraint of discrete variables into continuous variables. For example $x \in \{0,1\}^n \to x \in [0,1]^n$.

2: Obtain the optimal solution of the relaxed problem x^*_{relax}. Typically with continuous relaxation, the relaxed problem can be solved by some optimization technique for continuous variables, such as linear programming (LP).

3: If $f(x^*_{\text{relax}}) \geq f(\tilde{x})$, this problem is fathomed.

4: If $x^*_{\text{relax}} < f(\tilde{x})$ and $x^*_{\text{relax}} \in X$, this solution is a solution of the original problem, and we can update the incumbent solution.

5: Otherwise, divide the search space into several subsets, say $X_1, X_2, \cdots, X_m$. Typically it can be done by specifying the value of a component of the discrete decision variables. Generate subproblems

$$(P_i)\quad x^* = \arg\min_{x \in X_i} f(x)$$

6: For all the subproblems, apply this procedure recursively, *i.e.* $BAB(X_i, f(x), \tilde{x})$.

For simplicity, the above prototype algorithm was described as the "depth-first search" algorithm for traversing the search tree. However, the search order is an important strategic matter in BAB, and "breadth-first search" or "search prioritized by lower bound" can be considered.

13.2 Tabu Search

For many combinatorial optimization problems, the computational cost of an exact solution grows exponentially with the dimensionality of the problem. For such cases, we need to resort to approximate solutions in practice, and the development of such solution methods have received much attention. Among these, besides the well-known *simulated annealing (SA)* method of Kirkpatrick *et al.* [62], another generic procedure is the *tabu search (TS)* introduced by Glover [174]. It is based on recording the progress of a local search procedure, and using that information to help in continuing the search from local optima. Because of its simple structure and wide applicability, we review here the TS method.

13.2.1 Definition of the Problem

Here we are concerned with the following optimization problem. The feasible solutions are in a finite set F, and there is a function E defined over elements of F. We need to find an $x \in F$ that minimizes E:

$$\min_{x \in F} E(x) \qquad (13.1)$$

We call F the solution (or feasible) space, and after an analogy from physics, E is sometimes called the energy.

Problem 13.1 is solved in principle in finite time, by enumerating all elements of F and selecting the best one; however, usually this is not feasible. For instance, if the solution space is n binary variables, its size is $|F| = 2^n$, which grows exponentially with n, and becomes infeasible even for a value of $n = 30$. Furthermore, even if we were to try to design an efficient search algorithm, it is conjectured that for the class of problems called *NP-hard*, there is no algorithm with nonexponential (polynomial) complexity in n. For this reason, there is a need for efficient approximate solution algorithms.

13.2.2 Local Search

As a preliminary to introducing TS, we need to define *local search (LS)*. In local search, we take an incumbent solution $x \in F$, and perturb it to get $x' \in F$, where the set covered by x' is written as $N(x)$, and called the *neighborhood of x*.

As an example, for a problem with n binary (0-1) decision variables, if we select one of the n variables and flip its value between 0 and 1, the neighborhood $N(x)$ is determined only by the choice of the perturbed variable, so its size is $|N(x)| = n$.

As another example, consider the *assignment problem*, where we need to determine the one-to-one assignment of the elements of a set A to the elements of another set B, with both sets having the same dimensionality n. If we define the perturbation of the current assignment as the selection of two pairs of assigned elements, and the exchange of this two assignments, the size of the neighborhood becomes $|N(x)| = \frac{n(n-1)}{2}$.

LS is obtained by selecting the perturbed solution x' as the new incumbent solution, if its objective function value is better than that of the current x; and repeating this process. LS stops when there is no better value in $N(x)$. Such a situation is called a *local optimum*.

We can write the LS algorithm as follows.

Algorithm 13.2 The prototype local search algorithm

1: Generate an initial solution x^0.
2: Set the iteration counter t to 0.
3: If we find that

$$\forall x \in N(x^t), E(x) >= E(x^t)$$

we return x^t as the approximate optimum, and stop.
4: Search for a solution in $N(x^t)$ that improves on x^t, and set x^{t+1} to that.
5: Set $t \leftarrow t + 1$ and go to **Step 3**.

Figure 13.1 illustrates the process of a LS. We assume that the neighborhood of x consists of its two neighbors: $N(x) = \{x - 1, x + 1\}$, and start the search from x^0. LS will continue updating the solution to the neighbor that has a lower function value, and eventually arrive at x^3. Since the neighborhood is the two neighbors, when there is no improvement in that range, the search stops. Thus, if we start the search from x^0, it is not possible to arrive at the optimal solution x^*.

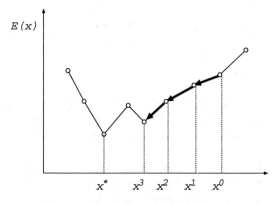

Fig. 13.1. Local search

In LS, by restricting the search to $N(x)$, we have achieved better efficiency, but as we have seen above, the search can stop at a local optimum. We can try to improve LS in several ways:

1. Find a better neighborhood definition.
2. Repeat LS starting from a different solution point. This is called *multistart local search (MSLS)*.
3. During LS, allow updating the solution not only to solutions with better values, but sometimes, subject to some restrictions, to worse solutions, too. This could allow the search to escape from local optima.

TS is one method of implementing the above 3rd solution.

13.2.3 Basic Structure of Tabu Search

The main point to understand about tabu search (TS), is how the update rule of LS is modified.

Recording and Use of the Search History

The basic idea of TS is to select the best solution from the neighborhood $N(x)$ of the current solution x, where x itself is excluded from $N(x)$. This will allow the selection of solutions that are worse than the current one. However, without further precautions, this would lead to selecting the previous solution again in the next step, and thus to cycling behavior.

A solution to this problem is obtained by collecting the recently visited solutions into the *tabu list* L, and prohibiting the selection of solutions in L. In other words, we select the next incumbent solution x' as the best solution in the partial neighborhood of x: $N(x) \setminus (L \cup \{x\})$, instead of in $N(x)$. This will prevent short (shorter than $|L|$) cycling.

By adopting such a modification to the updating rule, TS can avoid getting trapped at local optima, and can find global optima.

A Prototype TS Algorithm

The following is the outline of a TS algorithm:

The above prototype can be made more sophisticated based on various strategies, some of which are briefly listed here. The length of tabu list can be fixed (the number 7 seems to be popular), or variable (*e.g.* determined by a random number). The *tabu tenure* during which the accepted solution is forbidden to leave the tabu list can also be fixed or variable. How to determine the neighborhood of an incumbent solution is very important for the tabu search to be effective. Small neighborhoods are preferable from the viewpoint of the speed at which the local search proceeds. If the neighborhood is too small, however, the final solution may be of poor quality. Better solutions

---

Algorithm 13.3 A prototype tabu search algorithm

1: Generate an initial solution x and set it to the incumbent solution x^*.
2: Set the tabu list empty: $L = \emptyset$
3: If $N(x) \setminus (L \cup \{x\}) = \emptyset$, go to **Step 8**.
4: Select the best x' from $N(x) \setminus (L \cup \{x\})$, record the current solution as $x_{tabu} = x$, and update the current solution to $x = x'$.
5: If $E(x) < E(x')$, update the incumbent solution $x^* = x$.
6: If a prearranged termination condition is met, go to **Step 8**.
7: If the length of the tabu list is below a predetermined limit, add x_{tabu} to the tabu list. If the length is already at the limit, remove the oldest entry, before adding x_{tabu}. After that, go to **Step 3**.
8: Return the incumbent solution x^* as the resulting approximate optimum, and terminate the search.

---

can be obtained by using a larger neighborhood, but the computation time may not be acceptable. The diversification strategy is used in tabu search to enhance the search process by examining unvisited regions and by generating solutions that differ in a significant way from those obtained so far. An example of the diversification is the multi-stage tabu search, where different neighborhood structures are employed along the search process. An important element of flexibility in tabu search is the introduction of tabu solution to an admissible neighborhood by means of an aspiration criterion. The tabu status can be changed if a certain condition is satisfied, *e.g.* a tabu solution which is better than any previously obtained solution deserves to be considered.

We shall see an application of the TS algorithm to a transportation problem in an automated warehouse in Chapter 20.

Part IV

Topics in Modern Control for
Transportation Systems

14

The S-ring: a Transportation System Model for Benchmarking

In this chapter, we are going to capture the essence of a transportation system in a most simple model. We choose the elevator system as our target to be modeled. The goal will be to arrive at a system that cannot be reduced any further, but that still retains some interesting properties of the elevator system: its inherent instability, and the nontriviality of the optimal control law. By creating such a minimal model, we hope to gain a powerful tool for testing control algorithms, and to find direct insight into the dynamics of such complex systems.

We shall start from two directions:

- From the elevator system, we proceed by coalescing the states of the elevator controller into a minimal set, and at the same time we also eliminate the huge number of parameters that would define a given building.
- From the cellular automaton called the "Kac ring", which is introduced below, we proceed by generalization, adding a few new features.

The surprising thing is that the two approaches meet halfway. We call the result the "S-ring" [1], as this model is really a variant of the Kac ring.

Furthermore, the S-ring will be proposed as a benchmark for the comparison of different optimization algorithms in Chapter 16. It enables comparisons of algorithms such as Nelder-Mead simplex methods, quasi-Newton strategies, evolutionary algorithms, and particle swarm optimization.

14.1 The Kac Ring

The Kac ring, introduced by Marc Kac [175], is a deterministic discrete-state dynamical system that has the property that the trajectory of an instance of the system approximates the ensemble distribution. The system behavior

---

[1] "S" stands for "Sequential", as the S-ring evolves step-by-step, as opposed to the parallel evolution of the Kac ring

is thus seemingly probabilistic, in spite of the system being deterministic, finite-state, and time-reversible.

The Kac ring has been used to demonstrate some properties of systems of statistical mechanics. Here, we are concerned only with the most elementary results that are relevant to the S-ring that we want to build.

A particular Kac ring consists of n balls arranged in a ring, each being in either the "dark" or the "light" state. There are also m markers, placed at m positions around the ring, between adjacent balls. The evolution of the system is synchronous at each ball position. At each step, the balls move one position ahead clockwise. When a ball passes a marker, the state of the ball is reversed from dark to light and *vice versa*. We describe the state of the Kac ring by the number N of dark balls.

In Fig. 14.1, we show a few steps of the evolution of a Kac ring with $n = 8, m = 5$. We can see that N fluctuates wildly and seemingly randomly.

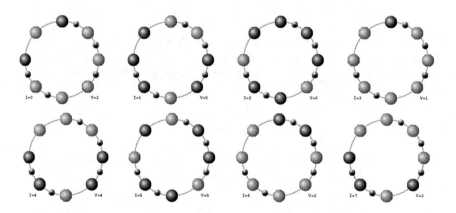

Fig. 14.1. Evolution of a Kac ring model

The ensemble distribution of N for all Kac rings of order n is easily calculated:

$$P(N) = 2^{-n} C_n^N$$

In Fig. 14.2, the empirical state distribution $\widehat{P}(N)$, obtained by running a single Kac ring, is compared with the theoretical probability distribution. Although a given Kac ring can take on only a small subset of the possible states, $\widehat{P}(N)$ approximates the ideal $P(N)$ reasonably well.

The S-ring introduced below shares with the Kac ring this property of having an interesting behavior in spite of the simple structure. It differs mainly by adding two features: stochastic state change, and external control.

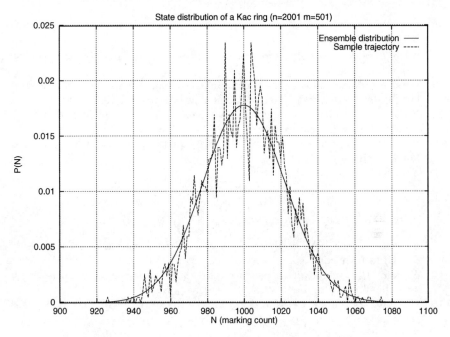

Fig. 14.2. An example of the state distribution of a Kac ring

14.2 Definition of the S-ring Model

We proceed to derive the model by stepwise elimination of some aspects of the elevator system. First, let us note that conceptually the elevators can be viewed as if running in a loop: they go up, then down, then up again,... The model is based on this observation, but to make it precise, we need to explicitly define the differences between real elevators and the model.

1. Model elevators will stop for taking on passengers, but not for discharging them.
2. Model elevators have unlimited capacity.
3. We don't keep account of boarded passengers.
   From 1 − 3, model elevators become indistinguishable and interchangeable.
4. Running direction is reversed only at terminal floors.
5. Passenger arrival rates are taken to be identical at all floors and directions, and denoted by p.
6. Floor distances are defined to be uniform.
7. We disallow cars passing each other: if the next floor is occupied by another car, this car cannot move forward until the next floor becomes empty.
   From 4 − 7, the system acquires a cyclic symmetry, with indistinguishable floors.
8. The states of the floors are represented by one bit $c_i (i \in \{0, \ldots, n-1\})$ each: $c_i = 1$ if one or more passengers are waiting, 0 otherwise.

9. Similarly, the cars are represented by one bit s_i for each floor: $s_i = 1$ if a car is present, 0 otherwise.
   With 8 and 9, the system state at time t is defined by the vector
   $x(t) = \{c_i; s_i\} \in X$.
10. Elevator running and stopping times are taken as uniform; both are unit length.
11. Time is discretized, events happening at unit intervals $t \in \{0, 1, \ldots\}$.
12. State transitions happen sequentially at the n sites, scanning them from $n-1$ down to 0, then around again.
    With 10 − 12, the system evolution becomes a discrete vector-valued Markov chain.
13. We define an objective function

$$h = n - \sum_{0}^{n-1} c_i \tag{14.1}$$

   a one-step reward $r(t) = h(t+1) - h(t)$, and the performance measure $H = E(h(t))$ to be maximized.
14. We equip the system with a policy $\pi : X \mapsto \{0, 1\}$, to generate decisions, with the optimal policy $\pi*$ defined as $\pi* = \arg\max_{\pi} H(\pi)$.
15. The state transitions are defined in terms of the state bits and the policy (see Table 14.1).

The above steps 1 − 15 complete the reduction of the elevator system to a CA (cellular automaton) [176]; which still can be seen in the terms of the elevator system (Fig. 14.3). If we look at the structure of this CA (Fig. 14.4d),

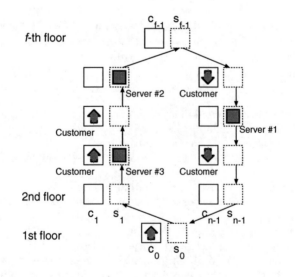

Fig. 14.3. The S-ring as an elevator system

Table 14.1. State transitions of the S-ring. The entries can be read as in the following example: The server has to make a decision (to "take" or to "pass" the customer) if there is a customer waiting (1xx), and if there is a server present on the same floor (11x) but no server on the next floor (110). It is obvious that there are states where no decision is required, resp., possible.

$\xi(t)$	Prob	$\pi(x)$	$\xi(t+1)$	Δr
000	$1-p$		000	0
	p		100	-1
100	1		100	0
010	$1-p$		001	0
	p	0	101	-1
		1	010	0
110	1	0	101	0
		1	010	$+1$
001	$1-p$		001	0
	p		101	-1
101	1		101	0
011	1		011	0
111	1		011	$+1$

the similarity with the Kac ring (Fig. 14.4c) is striking. However, the important difference lies in the presence of the policy and performance measure. These will allow us to use our system as a testbed for developing optimization algorithms for transportation systems.

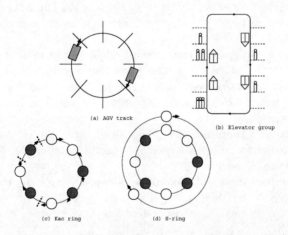

(a) AGV track

(b) Elevator group

(c) Kac ring

(d) S-ring

Fig. 14.4. The S-ring model and related systems

14.3 Control of the S-ring

The S-ring can be used to define an optimal control problem, by equipping it with an objective function H (here E is the expectation operator):

$$H = E(n - \sum c_i) \tag{14.2}$$

For given parameters n, m, and p, the system evolution depends only on the policy π, thus we can write $H = H(\pi)$. We define the optimal policy as

$$\pi^* = \arg\max_{\pi} H(\pi) \tag{14.3}$$

The basic optimal control problem is to find π^* for given parameters n, m, and p.

14.3.1 Representations of the Policy

In a general case, π can be realized as a lookup table of the state $\mathbf{x}$. Here, we assume that $\mathbf{x}$ is normalized by shifting it to start at the site where the decision is taken. An enumeration of $\mathbf{x}$ is defined by first regarding the state bits as the binary representation of an integer:

$$\chi = \sum_{i=0}^{n-1} 2^i (c_i + 2^n s_i) \tag{14.4}$$

Based on this, a compact index κ can now be defined as the ordinal number of the state when ordered according to χ. Here we are counting only the valid states, i.e. those satisfying $\sum s_i = m$. This index κ is used throughout the following.

In principle π^* is found by enumerating all possible π and selecting the one with the highest H. Since this count grows exponentially with n, the naive approach would not work even for the smallest cases; but we shall see that solution by dynamic programming is possible for small n, using the lookup table representation.

Another tabular representation is obtained by using a "Q-table", i.e., a matrix of the expected returns for each decision in each state. This representation is used for Q-learning.

A more compact representation of the policy is a linear discriminator (perceptron)

$$\pi(\mathbf{x}) = \theta(\mathbf{w}^T \cdot \mathbf{x}) \tag{14.5}$$

where θ is the Heaviside function. In general, there is no guarantee that the 0 and 1 classes of the optimal policy are linearly separable. However, we have found that for the case studied by us, the perceptron can realize the optimal policy.

Further possibilities include using a neural network, CMAC, decision trees, etc.

14.3.2 Policy Examples

The most obvious heuristic policy is the "greedy" one: when given the choice, always serve the customer:

$$\pi^g \equiv 1$$

Rather counterintuitively, this policy is not optimal, except in the heavy traffic ($p > 0.5$) case. This means that a good policy must bypass some customers occasionally.

Another trivial possibility is the "random" policy:

$$\pi^r = \begin{cases} 0 & w.pr.(1-\sigma) \\ 1 & w.pr.(\sigma) \end{cases}$$

This policy is rather poor, as expected.

A quite good heuristic policy is the "balance" policy:

$$\pi^b = \begin{cases} 0 \ if \ s_{n-1} = 1 \\ 1 \ otherwise \end{cases}$$

The intent is to put some distance between servers by 'passing' when there is another tailing server, letting it serve the customer. By balancing the positioning of the servers, π^b is significantly better than π^g for medium values of p.

We should note that even for a small S-ring neither of the above is an optimal policy; the real π^* is not obvious, and its difference from heuristic suboptimal policies is nontrivial.

14.3.3 Extensions

It should be noted that the basic problem discussed here could be extended in various ways:

- Find π^* when p is unknown (to be estimated), or for a range of p values.
- Find the best π for a given finite number of function evaluations.
- Find the best π when it belongs to a restricted class: *e.g.* only a finite window of $\mathbf{x}$ is available as information for decisions, or π is realized with a neural network with some predetermined parameters.
- Change the objective function H from expectation to a discounted infinite sum, or to a finite-horizon average.

It is also possible to modify the S-ring to correspond more closely to reality; by *e.g.* allowing nonstationary (time-dependent) $p = p(t)$, or allowing slightly more complicated dynamics. Such extensions are not considered here.

14.4 A Prototype S-ring

In the remainder of this book, (n, m, p)-sring will denote the S-ring with n sites, m elevator cars, and arrival probability p. We use the following model throughout if not stated otherwise: $n = 6$, $m = 2$, $p \in (0.1 \dots 0.9)$. Example evolution paths are shown in Fig. 14.5.

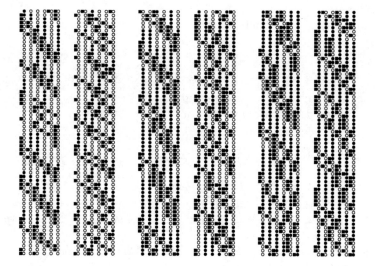

Fig. 14.5. Evolution path samples of the $(6, 3, p)$ S-ring, with the *greedy* and the *balance* policy for $p = 0.1$, $p = 0.3$, $p = 0.5$

In the figure, time runs downward, and only every nth step is shown. The rectangles are the servers, the circles the customers (black if present).

We can make some qualitative observations from the samples. The effect of increasing p is to cause the servers run around slower. Also, the samples show the tendency of the "greedy" algorithm to have the servers bunch together, decreasing the performance. The "balance" algorithm tends to separate the servers, thus increasing the probability that a new customer has an approaching server nearby. This effect is most clear in the case of $p = 0.3$. The general behavior in this respect is thus quite similar to actual elevator systems.

14.5 Solution by Dynamic Programming

14.5.1 Formulation

The optimal control problem of the S-ring can be formulated as a dynamic programming problem [177].

With states numbered by κ, we define the state transition probability from state i to state j for a control decision $\alpha \in \{0, 1\}$ as p_{ij}^{α}. For the policy $\pi = [\alpha_{ij}]$, the state transition matrix is defined as $P^{\pi} \equiv [p_{ij}^{\alpha}]$. Here p_{ij}^{α} can be read out from the second column of Table 14.1.

Similarly, we can define the reward matrix $R^{\alpha} = [r_{ij}^{\alpha}]$, where the elements are the changes in the control objective after each transition, obtained as Δr in Table 14.1.

For a given policy π, we introduce the notation $P = [p_{ij}] = P^{\pi}$, $R = [r_{ij}] = R^{\pi}$, and define the immediate expected reward vector $\mathbf{q} = [q_i]$ by

$$q_i = \sum_j p_{ij} r_{ij}$$

With these, we also define the gain g and the value vector v_i, as the values satisfying

$$g + v_i = q_i + \sum p_{ij} v_j \qquad (14.6)$$
$$v_{n-1} = 0$$

It can be shown that $H(\pi) = g$; thus the optimal policy is that which maximizes g.

14.5.2 Solution

The exact solution can be found by Howard's policy iteration algorithm:

Algorithm 14.1 Solution of a dynamic programming problem by policy iteration

---

1: Set $\pi = \pi_0$ (arbitrary), $P := P^{\pi_0}$, $R := R^{\pi_0}$ /* */Initialize
2: Calculate the gain g and the value vector v_i from Equation (14.6)   /*
   */Value determination
3: For each state i, find the decision α that gives

$$q_i^{\alpha} + \sum p_{ij}^{\alpha} v_j = \max$$

   and collect them as the policy $\pi := [\alpha_{ij}]$.   /* */Policy improvement
4: If the policy has been the same for the last two iterations, stop. Otherwise set
   $P := P^{\pi}$, $R := R^{\pi}$, and return to **Step 2**.   /* */Termination check

---

Results for the prototype S-ring are shown in Table 14.2.

14.6 Solution by Numerical Methods

We compare the performance of numerical optimization methods in solving the S-ring control problem. For Q-learning [178], we use a lookup-table representation. In the other methods, we use a single-layer perceptron for control,

Table 14.2. Performance of the $(6, 2, p)$ S-ring (π^g: greedy policy)

p	$H(\pi^g)$	$H(\pi^*)$
0.1000	5.0497	5.1624
0.2000	4.0292	4.3153
0.3000	3.3779	3.6017
0.4000	2.9637	3.0541
0.5000	2.6537	2.6723
0.6000	2.4163	2.4163
0.7000	2.2423	2.2423
0.8000	2.1211	2.1211
0.9000	2.0427	2.0427

and search for a set of connection weights $\mathbf{w}$ that maximizes the objective function H.

14.6.1 Kiefer–Wolfowitz Stochastic Approximation

We define the sample performance of a simulation run of length N steps (or simulation time N), with weight vector $\mathbf{w}$, as $\Theta(N|\mathbf{w})$. Using this, the Kiefer–Wolfowitz stochastic approximation algorithm [179] used here is given as:

Algorithm 14.2 Optimization by stochastic approximation

1: Set $i := 1$, $\mathbf{w}_i = \mathbf{w}_0$ (arbitrary)
2: With a normally distributed random vector $\mathbf{r}_i$, let $\Delta\mathbf{w}_i = \delta_i\mathbf{r}_i$, $\mathbf{w}_i^+ = \mathbf{w}_i + \Delta\mathbf{w}_i$, $\mathbf{w}_i^- = \mathbf{w}_i - \Delta\mathbf{w}_i$
   Collect the samples $\Theta_i^+ = \Theta(N|\mathbf{w}_i^+)$ and $\Theta_i^- = \Theta(N|\mathbf{w}_i^-)$ from consecutive simulation runs.
3: Update the weight vector: $\mathbf{w}_{i+1} = \mathbf{w}_i + \alpha_i(\Theta_i^+ - \Theta_i^-)$
4: Set $i := i + 1$. If $i < n_{\text{iter}}$, go to **Step 2**; otherwise terminate with $\mathbf{w} = \mathbf{w}_i$.

In these tests we used the values $N = 10\,000$ and $n_{\text{iter}} = 100$. Theoretically, we should decrease gradually α_i and δ_i to satisfy the convergence criteria. However, for the small number of iterations here we have used fixed values of $\alpha = 0.1$, $\delta = 0.5$. Also, the perturbations could have been done sequentially for each component of $\mathbf{w}$, and the perturbation step could be based componentwise on sensitivity analysis; the effect of such improvements should be tested later.

14.6.2 Q-learning and Evolutionary Strategies

For each step of a simulation run of the S-ring, the observed reward r_t can be used to estimate the expected discounted reward for each possible decision in each state, in a reinforcement learning scheme. It can also be averaged over a simulation run, and used as a fitness value estimator, for an evolutionary optimization method.

We have tried both the Q-learning algorithm, as described in Chapter 10, and a simple (1+1) evolutionary strategy, as described in Chapter 16, to optimize the policy of the S-ring, using the above sampling method. The results are given in the next section.

14.6.3 Results of the Optimization Experiments

We have compared the performance of the controllers found by the KW (Kiefer–Wolfowitz stochastic approximation) and ES ((1+1)-evolutionary strategy) algorithms, for different values of p. For the comparison, we have calculated the gain, which is obtained by executing the value evaluation step Equation (14.6) of the dynamic programming calculation for each controller. We have set the control decision $[\alpha_{ij}]$ for each state to be equal to the decision of a perceptron with the weight vector $\mathbf{w}$ obtained by each algorithm. The results are summarized in Table 14.3. For comparison, we have included the performance of the greedy, balance, and optimal controllers.

The results show that the ES algorithm consistently outperforms the KW algorithm or at least gets the same or equivalent result. For $p = 0.3$, the ES algorithm even found a solution superior to the balance heuristic; on the other hand, the KW algorithm never outperformed the better one of the two heuristics. However, the real optimum was not achieved by either algorithm.

The Q-Learning algorithm found the same or better solutions like the ES algorithm, at the cost of many more function evaluations. Also, it was necessary to finetune the learning parameters to get this result, which would not be possible in standalone operation.

Overall, we get the fastest convergence with the ES(1+1) algorithm. Q-learning reaches slightly better results, but in the present (lookup-table based) form it cannot be scaled up, and needs manual tuning of the learning parameters.

14.7 Conclusions

In this chapter, we have found the S-ring helpful in selecting and evaluating optimization methods for dynamic controlled systems.

From the simple tests, we can conclude that the ES algorithm is promising for constructing an optimal controller for stochastic dynamical systems.

In later chapters, we will embark on finding an ES variant that can reduce N and n_{iter}, and can get optimal or near-optimal results for a wide range of parameters. A new method, "Threshold Selection" [102, 180] will be seen to be superior for optimizing the control of not only the S-ring, but other related systems too.

Table 14.3. Performance results for each algorithm

p	$H(\pi^g)$	$H(\pi^b)$	$H(\pi^*)$	$H(\pi^{ES})$	$H(\pi^{KW})$	$H(\pi^{QL})$
0.1000	5.0497	5.1465	5.1624	5.1209	5.0497	5.1213
0.2000	4.0292	4.2787	4.3153	4.2211	4.0292	4.2767
0.3000	3.3779	3.5688	3.6017	3.5843	3.4968	3.5882
0.4000	2.9637	3.0500	3.0541	2.9441	2.9637	2.9837
0.5000	2.6537	2.6722	2.6723	2.6516	2.6535	2.6258
0.6000	2.4163	2.3970	2.4163	2.4163	2.4155	2.3914
0.7000	2.2423	2.2031	2.2423	2.2423	2.2423	2.2234
0.8000	2.1211	2.0765	2.1211	2.1211	2.1209	2.1118
0.9000	2.0427	2.0091	2.0427	2.0427	2.0427	2.0389

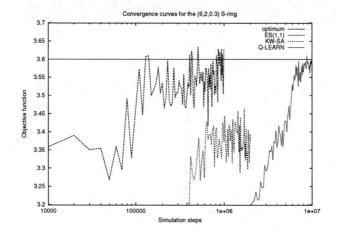

Fig. 14.6. Comparison of the convergence of the numerical methods

15

Elevator Group Control by Neural Networks and Stochastic Approximation

In this chapter we will consider a new elevator group control method, which uses neural networks to improve the performance and enhance the flexibility of elevator systems. The development of this method was based on the following reasons:

- New types of buildings, among them the so-called *intelligent buildings*, and changing economical conditions have caused more and more variations in building usage.
- The *fuzzy AI group control method* (see Section 9.3.2) was successful in many typical office buildings, for those traffic cases that were targeted originally. However, creating new fuzzy expert rules for all other, less-typical buildings and traffic conditions did not seem feasible. Also, the expert-system approach depends on information provided by human experts. Therefore, it cannot be used for automatic, onsite tracking of changing conditions, since we cannot provide each building with a human expert. There was thus a need for an improved group-control system, that could adapt itself to new target sites, or changing usage at existing sites, without the need for reprogramming.
- Since the main missing capability of previous methods was the learning ability, new research results made neural networks attractive to use as a general-purpose adaptive mapping mechanism in controllers. By applying and extending the neural-network technology, it was possible to create an adaptive algorithm for elevator group control.
- Because of the advances in computer technology, it became possible to use more computationally intensive methods, if they promised better performance. Even relatively high-speed floating-point processing and multiprocessing became feasible for installation in elevator controllers. This allowed the implementation of neural-network technology in elevator group controllers.

In the first part, after defining the elevator group control problem, the basic structure of the author's solution for controlling an elevator group with neural networks is described. In the following parts, we will see how an efficient group controller can be built with the neurocontroller, and discuss conditions that are specific to elevator systems. In the last part, a reinforcement learning method is introduced, that is applicable to the offline adaptation of the neural group controller, and that can also form the basis for online, onsite application.

15.1 The Elevator Group Control as an Optimal Control Problem

In the following, we consider only one version of elevator group control, called *immediate announcement*, which is most popular at present in Japan.

The controller operates by allocating one elevator car to each new hall call immediately. The allocation is announced to the waiting passengers, by lighting one hall lantern, corresponding to the direction of the call, over the entrance of the allocated car. This enables the passengers to queue up and board efficiently when the car arrives, instead of trying to guess which car will come, and rushing to the opening door when it finally arrives. After an allocation is announced, the controller tries to keep the promise, and the allocation is changed only when it becomes inevitable, *e.g.* because the selected car became extremely delayed, or it became fully loaded at another floor and had to bypass the current floor.

For the above class of elevator systems, the problem of optimal control of an elevator group can be stated in a simplified, finite-horizon form as follows.

Consider an elevator system, with the state vector $\xi(t)$ representing the car positions, traveling directions, waiting and traveling passengers, *etc.*, at time t. We denote the observable state vector seen by the controller by $\mathbf{x}(t)$, where the most significant difference between ξ and $\mathbf{x}$ is the absence of information about individual passengers in the latter.

We denote finite samples of traffic by

$$\Theta^p = \{(\tau_1^p; o_1^p; d_1^p), ..., (\tau_N^p; o_N^p; d_N^p)\}$$

where N is the number of passengers, and p, the index of the sample set, will usually be omitted. Here τ_i $(i = 1, \ldots, N)$ are the passenger arrival times, with the interarrival times $\Delta\tau_i = \tau_i - \tau_{i-1}$ distributed randomly, and o_i and d_i are the origin and the destination floors of the passengers, also distributed randomly.

Hall calls $a_k \in \{1, \ldots, N\}$ $(k = 1, \ldots, n)$ are generated by those passengers who arrive at the head of a hall waiting queue. Here $n \leq N$ is the number of hall calls, which depends not only on the traffic Θ, but also on the control decisions, introduced below.

The controller allocates each hall call to an elevator car when the call is registered; namely the decision variable at $t_k = \tau_{a_k}$ is the allocated car number $c_k \in \{1, ..., m\}$, where m is the number of the cars. Generally, c_k is given as a function ψ of the cumulative system state $\mathbf{X}_k$, with the control policy g as parameter:

$$c_k = \psi(\mathbf{X}_k, g) \qquad (15.1)$$

where $(\mathbf{X}_k)^T \equiv (\mathbf{X}(t_k))^T = [\mathbf{x}^T(t_0), ..., \mathbf{x}^T(t_k)]$ is the set of state vectors up to t_k. Here we will consider only the case where c_k depends explicitly only on $\mathbf{x}^T(t_k)$, not on the whole $\mathbf{X}_k$. In making this restriction, we assume that components of $\mathbf{x}$ contain the necessary dynamical information, *e.g.* counter values and memories, which are sufficient for control, without carrying along the whole cumulative state information $\mathbf{X}_k$.

The calls are served by car c_k at some later time θ_k, with the hall-call waiting times $h_k = \theta_k - t_k$ depending on the elevator dynamics, the calls, and also on the control decisions. On the basis of h_k, we can define an objective function. Ideally, the average passenger waiting time would be the most obvious choice as the objective function to be minimized. However, the observable information is not the passenger waiting time but the hall-call waiting time, hence we use here the average hall-call waiting time instead:

$$H = n^{-1} \sum_{k=1}^{n} h_k \qquad (15.2)$$

Thus, the optimal policy g^* is the one to minimize the objective function:

$$H(g^*) = \min_{g \in G}(H(g)) \qquad (15.3)$$

where G is the set of admissible policies, which depends on the structure of the controller and on several operating constraints.

It should be noted that the objective function of Equation (15.2) is not the only possible one. If the passenger arrival density could be estimated reliably, H could be corrected to approach the average waiting time of the passengers, instead of the average waiting time of the hall calls, *i.e.* of the first passengers in the queue. The selection of the average waiting time as the primary objective function also implies a value judgment, that the service level is thought to be more important than other performance indicators, *e.g.* as compared with the total system throughput. Further research will be needed to investigate the effect of other choices of the objective function.

15.2 Elevator Group Control by Neural Networks

When considering the construction of a controller for the elevator group-control system, a few important points should be noted.

- Control decisions are made only at discrete times, when a new call is registered. This is in contrast with those control methods that allow re-allocations at any time, so that control decisions have to be made continuously, *e.g.* "shall we change one allocation now or wait a little longer?"

- Since changes of allocations are very undesirable, each decision is usually final and any mistake has serious consequences. Therefore, an efficient controller would need to predict the future with high reliability, to avoid costly mistakes. A crucial part of the controller is the handling of the prediction of car movements, using all available information.

The requirement of the controller having built-in predictive capability will be taken into account when introducing the neurocontroller.

The operation of the elevator group control system, ignoring the possibility of allocation changes, can be stated formally as follows. At the time t_k of the kth new hall call, for a given observable state $\mathbf{x}_k$ of the elevator system, and for a given control policy g, find the allocation

$$c_k = \Psi(\mathbf{x}_k, g) \tag{15.4}$$

The above Equation (15.4) suggests that the controller can be realized as a mapping from the system state $\mathbf{x}$ into the decision c, parameterized with the control policy g. This observation forms the basis for introducing a neural network as the core part of the controller. The state vector $\mathbf{x}$ can then be used as the input signal to the neural network, and c is obtained by a simple postprocessing of the output signals; g is represented by the connection weights of the neural network.

In the following, we define the state vector $\mathbf{x}$. In the next section, we introduce the neurocontroller that represents the policy g.

15.2.1 State Representation for Elevator Group Control

We can define an observable state vector $\mathbf{x}_k$ that contains dynamic and static (constant) information about the elevator system in some form. Dynamic information, directly obtainable from the elevator system, might include the following:

1. *for elevator cars*
   a) operational state (automatic, attendant, manual, shut-down, *etc.*)
   b) control state (running, stopped, *etc.*)
   c) position (floor, distance from next floor) †
   d) direction
   e) allocated hall calls
   f) registered car calls
   g) load (passenger count)
2. *for the whole system*
   a) registered hall calls in the group
   b) elapsed waiting times for the calls †
   c) estimated traffic densities between all floors †
   d) estimated trend of traffic change †

In addition, static information is also necessary:

1. *for elevator cars*
   a) rated load (number of passengers)
   b) contract speed
   c) served floors
   d) floor distances †
   e) floor-to-floor running times †
2. *for the whole system*
   a) number of cars
   b) usage patterns of the building †

Furthermore, there are nonobservable state variables:

1. *for elevator cars*
   a) number of persons going to each floor
2. *for the whole system*
   a) number of persons waiting to go to each floor

In the above state variables, those marked with † are continuous, the others are discrete; so by combining them, $\mathbf{x}$ itself will not be discrete, and therefore cannot be represented as the index to a table. If the continuous parameters are discretized, $\mathbf{x}$ can be made discrete. However, if it were used as the index to a table, to implement a lookup-table-based method, that table would still have a very large total size, making it practically unusable.

For comparison, the approach taken by Levy and coworkers [55, 56] is to reduce $\mathbf{x}$ into a low-dimensional discrete space $\mathbf{b}$, by ignoring a large part of the information, and by reducing the remaining (binary) state variables through boolean simplification. Since in Levy's method the decisions are about immediate actions (whether a car should stop at a floor to serve a call, or pass it), the prediction of future movements is not directly needed, and therefore such a crude state descriptor is sufficient.

We take a different approach, and define a number of continuous-valued state variables $\mathbf{u}$, which compress the elevator system state into meaningful variables. This approach is based on the observation that an existing elevator group controller, using fuzzy-rule-based control, embodies a large amount of heuristic knowledge, in the form of the selection of input signals and the fuzzy rule-base. The new state indicators are selected by *reusing the expert knowledge*, which is embedded in the fuzzy AI group controller.

In the reduction of the state vector, our premises are

1. Most of the resulting state can be a "snapshot" taken at the time of hall-call allocation, extended with statistical information about the traffic and global system state. The elevator dynamics is assumed to be known, making it unnecessary to infer it from state change sequences. Therefore, detailed state information from previous instants is eliminated, except for the elapsed waiting times of hall calls, and arrival probabilities.

2. All possible symmetries (translational, permutational) are exploited and
relative information is used wherever necessary. In particular, call information is mostly useful when taken relative to the car position, like, e.g.,
"the number of existing calls between the present position of the car and
the floor of the new call." Although elevator states shifted vertically are
not strictly invariant, there is a significant equivalent component, which
is separated out by the heuristic state variables.

For the neurocontroller, the state vector $\mathbf{u}$ is built up from the following types
of components:

1. input variables of the fuzzy controller
2. output variables of the fuzzy controller
3. additional state indicators

The above components are organized as r input components for each of the
m cars, for a total of $m \times r$ state vector components:

$$\mathbf{u} \equiv [\mathbf{u}^{(1)}, \dots, \mathbf{u}^{(m)}] \tag{15.5}$$

Many of the *input variables* depend on an estimated (heuristic) value of the
arrival time θ_i^k of each car k to each floor i. The arrival time is conditioned
on whether the new hall call is allocated to that car or not. From θ_i^k the
estimated waiting times of the calls h_i^k are directly obtained. This allows the
construction of various merits for or against the allocation of the new call to a
car; *e.g.* the longest increase in expected waiting time for other calls; balance
of waiting times *etc.*

Other merit values consider the coincidence of hall calls and car calls, which
is important for reducing the number of stops, and avoiding the situation
where another car stops on a floor before the allocated car arrives.

It is also important to consider some extreme events, like, *e.g.*, the possibility of a car bypassing an allocated hall call, or leaving passengers at a floor,
because of a full load.

The *output variables* represent combinations of the merit values, while incorporating some expert rules (see Chapter 1). The output variables of the
fuzzy controller show when one of these rules is triggered by the system state,
and the degree that the rule applies to each car. The resulting *fuzzy member-
ship values* indicate for each car whether it should be a candidate for allocation
or not according to that rule. In addition, we also use directly the allocation
result of the fuzzy controller as one additional input variable for each car.

Furthermore, we might augment the state indicators by sensor and communication information (*e.g.* congestion sensors at selected floors, reservations
for elevator service at given floors and times, *etc.*). If we use this information
as the input to neural networks, it is not necessary to explicitly find relations
between the various types of state information. This is an advantage for the
designer, when compared with the fuzzy controller, where rules have to be

created for each new input. If the added information is useful for the neuro-controller, it will be used automatically. On the other hand, if the included information is superfluous, the neurocontroller will also automatically ignore the useless input as noise, and the only loss is that the neural network becomes somewhat larger and more difficult to train.

Table 15.1 shows an example of the state vector that was used for the neurocontroller. Here the number of state-vector components r was 25, which is the number also used in later sections. Besides this, we have also done tests both with fewer (starting with $r = 5$), and more ($r = 36$) components, and $r = 25$ was obtained as a compromise between slightly better performance and larger network sizes.

15.3 Neurocontroller for Group Control

In this section a neurocontroller is introduced as the controller of a stochastic system. The overall strategy is to construct first a controller that can control a plant with acceptable performance. It can then be given the job of controlling the actual, physical plant, and let it improve itself through "on-the-job train-ing". This strategy is implemented by emulating the control policy of a proven, existing controller; then starting an adaptation process for the control policy from that initial state. Supervised training is used for the emulation (initializa-tion), and reinforcement learning for the adaptation. One novel feature of the present method is the combination of supervised training and reinforcement learning for the same neural network, which is used as the trainable/adaptive element of the controller.

The controller is constructed in the following steps (see Fig. 15.1):

1. Supervised learning is used to embed domain-specific knowledge in the controller, by training a neural network to perform a prediction task re-lated to the control problem. At this stage, the plant is controlled by an existing conventional controller (the "prototype controller" in Fig. 15.1).
2. The controller is initialized by again using the supervised learning, to emulate the control policy of the prototype controller. At this stage, part of the information obtained in the previous stage is retained, in the form of the connection weights $\mathbf{w}^1$ of the hidden layer of the neurocontroller.
3. In the final phase the neurocontroller replaces the prototype controller, and a reinforcement learning is used to optimize the performance. Perfor-mance fluctuations are kept within a permissible range by using a sensi-tivity analysis, to allow an online adaptation.

The neurocontroller is explained in details only for the case of elevator group control. However, it should be fairly straightforward to apply the same method to other stochastic control problems, too.

Table 15.1. Input signals at the time of allocation (for each car)

u_1	Number of hall calls
u_2	Variance of distances between cars
u_3	Running average of u_2
u_4	Running average of u_1
u_5	Running average of UP hall-call rate
u_6	Running average of DOWN hall-call rate
u_7	Running average of number of car calls (from main terminal)
u_8	Running average of number of car calls (to main terminal)
u_9	Running average of number of car calls (other floors)
u_{10}	Allocated hall calls nearer than new hall call
u_{11}	Allocated hall calls beyond new hall call
u_{12}	Car calls nearer than new hall call
u_{13}	Car calls beyond new hall call
u_{14}	Passengers in car (load)
u_{15}	Free-car indicator
u_{16}	Longest route to new hall call
u_{17}	Distance to nearest other car
u_{18}	Car position (including direction)
u_{19}	Shortest route to new hall call
u_{20}	Estimated number of future hall calls before reaching new hall call
u_{21}	Estimated waiting-time increment (for new hall call)
u_{22}	Estimated worst-case waiting time
u_{23}	Estimated increase in bunching
u_{24}	Estimated waiting-time increment (for future calls)
u_{25}	Estimated waiting time

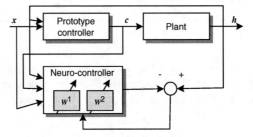

Step 1. Learn to predict the hall call waiting time

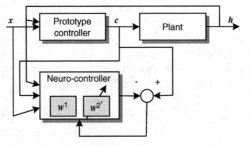

Step 2. Learn the control policy of the prototype controller

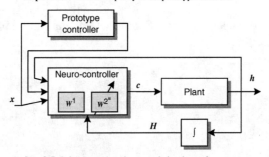

Step 3. Reinforcement learning to optimize the performance

x: system state w^1: connection weights to the
c: control decisions   hidden layer
h: hall call waiting times w^2: connection weights to the
H: performance measure   output layer

Fig. 15.1. The steps of constructing a neurocontroller for elevator group control

15.3.1 Structure of the Neurocontroller for Elevator Group Control

In designing a neurocontroller, we have to decide the following points:

1. What *neural-network architecture* to use?
2. How to *map* output neuron *activations into decisions*?
3. How to take into account the *structure of the elevator group-control problem*?

In the following section we will show that the Multi-Layer Perceptron (MLP) network has several advantages for this task; therefore this architecture was selected.

One possible implementation of an elevator group controller by a MLP is described, *e.g.* in [181] and [182]. With this method, the control policy g is realized as a mapping from the system state to the control decision

$$g : \{\mathbf{u}_k\} \mapsto \{c_k\} \tag{15.6}$$

The (discrete) decisions c_k are obtained from the continuous-valued outputs of a MLP network in the following way. At each decision time t_k, the input layer of the MLP acquires the state of the elevator system $\mathbf{u}_k$, encoded as a vector normalized in $[0, 1)^{rm}$, where m is the number of cars, and r is the number of state variables for each car.

The MLP network has continuous-valued output variables. There are several possibilities for the mapping of m real values into $a \in \{1, \ldots, m\}$, as shown in Fig. 15.2.

In the following, scheme (a), *i.e.* selecting the car for which the output of the neural network is the maximum, is used. Although scheme (b) has the advantage that it can be used with other neural networks, *e.g.*, LVQ nets, which have already discretized outputs; however, in the case of MLP, scheme (a) is more general (it can take into account the relation among all cars simultaneously).

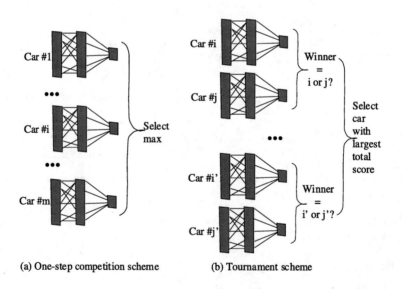

(a) One-step competition scheme   (b) Tournament scheme

Fig. 15.2. Two possible schemes for getting discrete decisions from continuous mappings

The transfer function between the input and the output layers is

$$[z_k^1, \ldots, z_k^m] \equiv \mathbf{z}_k = \mathbf{\Phi}(\mathbf{w}, \mathbf{u}_k) \tag{15.7}$$

where $\mathbf{w}$ contains the connection weights, viewed as a single vector.

The dimension of the output $\mathbf{z}$ of the MLP equals the number of cars, and the car corresponding to the largest output signal is allocated to the call:

$$c_k = \arg\max_\alpha z_k^\alpha \tag{15.8}$$

With this structure, the controller can be trained with allocation examples [181], which are obtained from, *e.g.*, simulation runs of a conventional group control system [59, 60].

An important further point is that the elevator system has a strong characteristic permutational symmetry. If the states $\mathbf{s}^{(i)}$ and $\mathbf{s}^{(j)}$ of any two cars i and j are interchanged, we want the decision to follow the permutation (if $c_k = i$ or $c_k = j$ then c_k is switched, otherwise c_k remains the same). If we write out explicitly the input state vectors of each car as

$$\mathbf{u}^T = \left[(\mathbf{u}^{(1)})^T, \ldots, (\mathbf{u}^{(m)})^T\right] \tag{15.9}$$

then the symmetry condition is described by

$$\mathbf{Z} = \begin{bmatrix} z^{i_1} \\ \vdots \\ z^{i_m} \end{bmatrix} = \mathbf{\Phi}\left(\mathbf{w}, \begin{bmatrix} \mathbf{u}^{(i_1)} \\ \vdots \\ \mathbf{u}^{(i_m)} \end{bmatrix}\right) \tag{15.10}$$

where $\{i_1, \ldots, i_m\}$ is an arbitrary permutation of $\{1, \ldots, m\}$ and $z^{i_1}, \ldots, z^{i_m}$

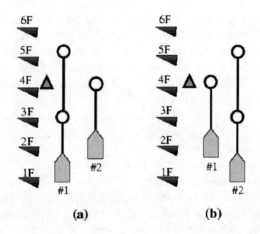

(a)   (b)

Fig. 15.3. Illustration of the permutational symmetry of the elevator system

are the network output values corresponding to car $i_1, \ldots,$ car i_m.

This symmetry might be satisfied implicitly by training with a sufficiently large training set. However, by explicitly enforcing the symmetry of the MLP by linked weights, we can get significant savings in the training time [181]. Assuming a three-layer architecture, Equation (15.7) can be expanded as

$$\mathbf{v}_k = \sigma(Q\mathbf{u}_k)$$
$$\mathbf{z}_k = \sigma(R\mathbf{v}_k) \qquad (15.11)$$

where $\sigma(\mathbf{x})$ is a vector-valued sigmoid function, with component values

$$\sigma(x) = \frac{1}{1 + e^{-x}} \qquad (15.12)$$

for each element of $\mathbf{x}$, and Q and R are the connection matrices. To satisfy the symmetry conditions of the elevator system, we write the activations of the hidden-layer neurons by

$$\mathbf{v}_k^T = \left[(\mathbf{v}_k^{(1)})^T, \dots, (\mathbf{v}_k^{(j)})^T, \dots, (\mathbf{v}_k^{(m)})^T \right] \qquad (15.13)$$

where each $\mathbf{v}_k^{(j)}$ has length q. Equation (15.11) can then be further expanded as

$$\mathbf{v}_k^{(i)} = \sigma(\sum_{j \neq i} B\mathbf{u}_k^{(j)} + A\mathbf{u}_k^{(i)}), \qquad i = 1, \dots, m$$
$$z_k^i = \sigma(\sum_{j \neq i} \mathbf{f}^T\mathbf{v}_k^{(j)} + \mathbf{e}^T\mathbf{v}_k^{(i)}), \qquad i = 1, \dots, m \qquad (15.14)$$

The dimensions of the submatrices A, B and row vectors $\mathbf{e}^T, \mathbf{f}^T$ are $(q \times r)$ and q, respectively, see Fig. 15.4 (left). Notice that A, B and $\mathbf{e}^T$, $\mathbf{f}^T$ in Equation (15.14) do not depend on i and j. This constraint is enforced during all phases of the learning and adaptation process, by linking the corresponding weights together. It is obvious that Equation (15.14) satisfies the conditions of symmetry given by Equation (15.10).

15.3.2 Initial Training of the Neurocontroller

An important consideration is how to ensure a sufficiently rich internal representation for the controller network. If there is strong correlation between some input signals and the training signal, there is a danger that the resulting network will be trivial. This could be the case, *e.g.*, when one of the input signals is the training signal (the allocation signal of the fuzzy controller) itself, like in the case of the state variables of Table 15.1.

For this reason, a two-stage training procedure [182] is adopted here. First, we pretrain the MLP with a sufficiently nontrivial task, using $\mathbf{u}_k$ for the inputs, and using the backpropagation learning. The pretraining task has to be able to give rich information to the network, and also needs to have a teacher signal η_k available for the backpropagation training. Since we are interested only in having a rich representation in the hidden layer, the output layer at this stage need not have the same structure as the controller network, see Fig. 15.4 (right).

The initial network learns the mapping

$$\Gamma : \{\mathbf{u}_k\} \mapsto \{\eta_k\} \tag{15.15}$$

Next, we replace the output layer and retrain the network with the allocation sample set. The training output vectors are constructed by observing the control decisions $\{\gamma_k\}$ generated by the fuzzy group controller, and the network is trained to perform the mapping

$$g^0 : \{\mathbf{u}_k\} \mapsto \{\mathbf{y}_k\} \tag{15.16}$$

where

$$y_k^j = \begin{cases} 1 \text{ if } j = \gamma_k \\ 0 \text{ otherwise} \end{cases} \tag{15.17}$$

During the learning, the connections up to the last-but-one layer are kept constant, hence even the "delta-rule" training procedure [183] is sufficient, see Fig. 15.4 (left). Since only the connections to the output layer R are changed, the internal representation of the hidden layers determined by the initial task Γ remains intact, and it is utilized for the decision making.

In the examples reported here, the prediction of the hall-call waiting times h_k was selected as the task Γ in the first training phase, by defining

$$\eta_k = 1 - \min(1, \frac{h_k}{h_{\max}}) \tag{15.18}$$

where $h_{\max} = 120$ [s] is a scaling factor. An example of the waiting-time prediction after the pretraining phase is shown in Fig. 15.5.

This choice of η_k has the advantage that the training signal is closely related to the system performance index H. To select the car with the lowest expected value of the waiting time is a plausible heuristic control policy. Therefore, a controller could be designed by calculating η_k for each car, and selecting the car corresponding to the largest value. For the pre-training, the input signals were reordered so the signals of the allocated car c_k came first:

$$\mathbf{u}^T = \left[(\mathbf{u}^{(c_k)})^T, \ldots\right] \tag{15.19}$$

It was found that we could obtain a good initial value for the output weight matrix R of the control network, by making m copies of the output weights of the predicting network. Figure 15.4 illustrates the correspondences. In effect, the neurocontroller at this stage predicts the η_k^i value for each car, and selects the car with the largest value (*i.e.* with the shortest predicted waiting time).

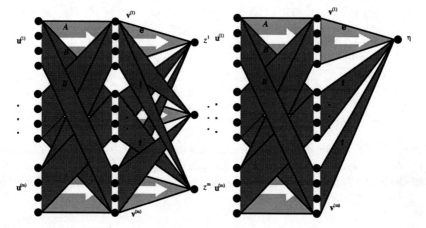

Fig. 15.4. *Left:* NN for control, *right:* NN for prediction. Both NNs with symmetry conditions.

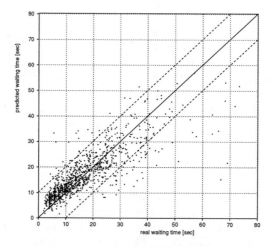

Fig. 15.5. Prediction of waiting times by a neural network

The prediction performance (the total error between the predicted and actual values of η_k) has been used to help the selection of effective input signals, by observing the effect of adding or changing an input signal on the prediction accuracy. The relevance of the inputs shown in Table 15.1 has been verified by checking that each u_i contributes to the prediction.

The training method for the supervised learning in Phase 1 and Phase 2 was a variant of the BP learning [184], based on a conjugate gradient descent optimization. This method, which was found fast and easy to use, was mentioned earlier in Chapter 10.

15.4 Adaptive Optimal Control by the Stochastic Approximation

The neurocontroller, when trained with samples from the fuzzy controller, is able to start controlling an elevator group at about the same performance level as the teacher system (see Table 15.4 for an example of performance at this stage). To improve further the performance, we can use the reinforcement training to tune the neural network of the controller.

Because of the requirement of online adaptation, and the stochastic nature of the plant, deterministic training procedures are not appropriate. That is, in an online situation we can make only noisy measurements of the system performance.

Since the MLP has continuous-valued parameters, it is possible to use the Kiefer-Wolfowitz Stochastic Approximation Algorithm (abbreviated below as KW-SAA) [185] as a reinforcement learning procedure [182]. With the KW-SAA algorithm, the system performance is sampled for perturbed values of the adjustable parameters, to perform a descent search toward the optimum. This means that there exist some intervals where the system is operated with perturbed controller parameters. Therefore it is necessary to have a good estimate of the parameter sensitivities in the case of the online adaptation, to avoid running the systems with unacceptably bad performance during those intervals.

One advantage of using the MLP for the controller is that the sensitivity of the output with respect to the network connections is easily estimated. In particular, the case of the connections to the output layer is especially simple; although the output neurons have nonlinear transfer functions, the sensitivity of the control decisions w.r.t. the connection weights is easily estimated by a linear relationship (see Section 15.4.2). This allows us to keep the performance fluctuations in a given range, by restricting the change in the control decisions between adaptation steps.

15.4.1 Outline of the Basic Adaptation Process

As noted above in Section 15.3, the connection weights of the neurocontroller are initialized by supervised learning, and during adaptation only the output weight matrix is modified. We further restrict the adaptation to the subvector **e** (see Equation (15.14)), since this can be shown to be sufficient for generality. Therefore, the νth adaptation step ($\nu = 1, ..., n_{iter}$) is given mathematically by

$$\mathbf{e}_\nu = \mathbf{e}_{\nu-1} - a_\nu \mathbf{I}^\nu \frac{H_\nu^+ - H_\nu^-}{s_\nu} \tag{15.20}$$

where the performance estimates for the perturbed weight vectors

$$H_\nu^+ \doteq H(\mathbf{e}_{\nu-1} + \mathbf{I}^\nu s_\nu)$$
$$H_\nu^- \doteq H(\mathbf{e}_{\nu-1} - \mathbf{I}^\nu s_\nu) \tag{15.21}$$

are obtained by running the elevator system for a given period with each perturbed neurocontroller. $\mathbf{I}^\nu$ is a cyclically shifted vector of $[1, 0, \ldots, 0]$, $\mathbf{e}_0$ is the initial weight vector learned by emulating a fuzzy controller [60]. An upper limit on s_ν is determined by setting the expected rate of allocations altered by the perturbation to a given number ε (for details of the determination of the perturbation range see Section 15.4.2 below). The detailed algorithm is as follows:

Algorithm 15.1 The overall structure of the adaptation algorithm for elevator group control

---

1: **Initialization:**
   Run the elevator system with the prototype (fuzzy) controller, using the traffic sample Θ^0. Collect samples $\{\mathbf{u}_k^0\}$, $\{c_k^0\}$, and $\{\eta_k^0\}$, and measure the initial performance H^0.
2: Pre-learn the connection weights Q and R of the neurocontroller by learning the sample set $\{(\mathbf{u}_k^0, \eta_k^0)\}$, using backpropagation [183] or similar methods [184].
3: Initialize the connection weights of the neurocontroller by re-using Q from Step 2, and learning the sample set $\{(\mathbf{u}_k^0, c_k^0)\}$ by the delta-rule.
4: **Adaptation:**
   Run the elevator system with the neurocontroller, using the traffic sample Θ^0. Estimate the smallest perturbations $\{\delta_0^\mu\}$ $(\mu \in (1, \ldots, q))$ according to Equation(15.28).
5: Initialize $\nu \leftarrow 1$, $\mu \leftarrow 1$, a_ν, $\hat{s}_\nu$.
6: Set $s_\nu \leftarrow \hat{s}_\nu \delta^\mu$.
7: Set $\mathbf{e}_\nu^+ \leftarrow \mathbf{e}_\nu + s_\nu \, \mathbf{I}^\mu$. Run the elevator system with $\mathbf{e}_\nu^+$, using a traffic sample $\Theta^{\nu+}$. Measure H_ν^+.
8: Set $\mathbf{e}_\nu^- \leftarrow \mathbf{e}_\nu - s_\nu \, \mathbf{I}^\mu$. Run the elevator system with $\mathbf{e}_\nu^-$, using a traffic sample $\Theta^{\nu-}$. Measure H_ν^-.
9: Update $\mathbf{e}_{\nu+1} \leftarrow \mathbf{e}_\nu + a_\nu \, \mathbf{I}^\mu \left(H_\nu^+ - H_\nu^- \right) / s_\nu$.
10: If $\nu < n_{iter}$, set $\mu \leftarrow (\nu \bmod q) + 1$, $\nu \leftarrow \nu + 1$, update a_ν, $\hat{s}_\nu$, and go to Step 6.
11: End.

The above is an online algorithm, if all $\{\Theta^{\nu+}\}$ and $\{\Theta^{\nu-}\}$ are different. In the offline case, $\Theta^{\nu+} \equiv \Theta^{\nu-} \equiv \Theta^\nu$ have been used.

In our tests we have used parameters in the following ranges: $a_\nu \in [0.1 \, \nu^{-1} \cdots 1.0 \, \nu^{-1}]$, $\hat{s}_\nu = [0.1 \, \nu^{-1/3} \cdots 1.0 \, \nu^{-1/3}]$, $\varepsilon = 0.1$. In general, the learning-rate parameters s_ν and a_ν are also required to satisfy the usual convergence conditions of the KW-SAA:

$$\sum_{\nu=1}^{\infty} a_\nu = \infty, \qquad \sum_{\nu=1}^{\infty} \left(\frac{a_\nu}{s_\nu} \right)^2 < \infty, \qquad \sum_{\nu=1}^{\infty} a_\nu s_\nu < \infty \qquad (15.22)$$

In this research, both an offline procedure, and an online procedure have been tested. In the former, repeated simulation runs with the same passenger arrivals are used to evaluate H_ν^+ and H_ν^-. Contrarily in the latter case, consecutive segments of one evolution path (a single long simulation run) are used for all performance evaluations. The offline procedure gives significantly faster convergence, and the results given in the next section have been obtained by that. The fast convergence of the offline procedure is attributed to the variance reduction in the repeated simulation runs, with the observation noise of $(H^+ - H^-)$ being much smaller than in the online procedure. This is well understandable, considering the correlation of the evolution paths in the case where runs are repeated with the same traffic and a slightly perturbed control policy. On the other hand, when the controller is trained by the off-line procedure, it does not experience the full scale of the stochastic variability of the passenger arrivals, and then there is a danger of adapting to a restricted subset of the possible system states. Online adaptation is going to be the subject of further study.

A global search procedure [186] has also been tried out, using resampling of previously tested points in the search space. However, that method was not found advantageous in this problem.

15.4.2 Sensitivity of the Controller Network

The sensitivity of the system performance H w.r.t. the controller parameters (connection weights) is estimated indirectly. We are going to find such perturbation values that are expected to change the system performance at most to a given degree. A central concept used here is the *probability of decision change* P_c: *i.e.* the percentage of decisions that become different with the perturbed controller, as compared with the original decisions of the non-perturbed controller. The sensitivity of H is obtained from an estimate of the dependence of P_c on the perturbation.

First, we obtain an estimated upper limit on the permissible rate of change in the control signal c. The system is run with the prototype controller, but a known number εn of the n control decisions are replaced by random decisions. From the observed deterioration of H, an allowable ε is determined. From Fig. 15.6, we have chosen $\varepsilon = 0.1$, by setting the allowable performance deterioration (arbitrarily) to 10%. Next, we find such perturbations to the connection weights, which can be expected to cause at most (εn) changes in $\{c_k\}$.

In the following, the controller network is assumed to be a simple feedforward network with one hidden layer, and we consider only perturbations to the weights between the hidden and the last layer.

With the symmetry conditions of Equation(15.11), it can be verified that it is redundant to consider perturbations to both $\mathbf{e}$ and $\mathbf{f}$, and hence we will consider only the case of $\mathbf{e}$.

For each system state $\mathbf{u}_k$, we can determine explicitly for a perturbation δ_k^μ to a given connection weight e^μ, whether it will cause the decision c_k to change or not. Since generally all outputs z_k^i change when $\mathbf{e}$ is perturbed, in most cases there will be a point when an output value z_k^i becomes larger than

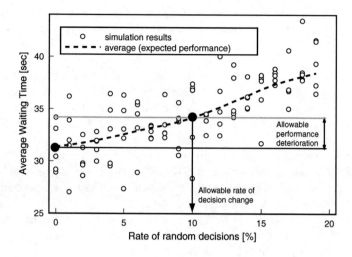

Fig. 15.6. Performance deterioration with random decisions. For each simulation run, out of $n = 300$ control decisions done by the prototype controller, (εn) are changed to random decisions.

$z_k^{c_k}$, thus causing the decision to change [1]. The point of equality (the *smallest* such perturbation that just causes the decision change) is the one used for the sensitivity analysis.

In the derivation below, we change slightly the notation, using $\mathbf{z}(t_k)$ instead of $\mathbf{z}_k$, and $z^i(t_k; \delta_k^\mu)$ for the perturbed value. Let us define the perturbed output equation by

$$z^i(t_k; \delta_k^\mu) = \sigma\Big(\sum_{j \neq i} \mathbf{f}^T \mathbf{v}_k^{(j)} + \mathbf{e}^T(\delta_k^\mu)\mathbf{v}_k^{(i)}\Big) \tag{15.23}$$

where

$$\mathbf{e}^T(\delta_k^\mu) = \mathbf{e} + \delta_k^\mu \mathbf{I}^\mu \tag{15.24}$$

and $\mu = (\nu \bmod q) + 1$ is the index of the perturbed element of $\mathbf{e}$ at the νth adaptation step.

To estimate the sensitivity of the kth control decision to the perturbation, we have to find the smallest δ_k^μ that yields output values $z^i(t_k)$ to satisfy

$$\exists i \neq c_k : z^i(t_k; \delta_k^\mu) = z^{c_k}(t_k; \delta^\mu(t_k)) \tag{15.25}$$

From Equation (15.23), and considering the monotonicity of the sigmoid function $\sigma(x)$, the equality condition of Equation (15.25) can be written as

---

[1] In order for c_k to change for a (positive or negative) δ^μ, either $z_k^{c_k}$ has to decrease, with at least one of the $(m-1)$ other outputs increasing, or $z_k^{c_k}$ might even increase, but with having one z_k^i that grows even faster. There is a good chance that one case is true at least for one direction of perturbation, and probably it will be true even for both positive and negative δ.

$$\sum_{j\neq i} \mathbf{f}^T \mathbf{v}^{(j)}(t_k) + \mathbf{e}^T(\delta_k^\mu)\mathbf{v}^{(i)}(t_k) = \sum_{j\neq c_k} \mathbf{f}^T \mathbf{v}^{(j)}(t_k) + \mathbf{e}^T(\delta_k^\mu)\mathbf{v}^{(c_k)}(t_k) \quad (15.26)$$

and after substitution from Equation (15.24) and simplification

$$\delta_k^\mu = \min_{i\neq c_k}\left[(v_\mu^{(i)}(t_k) - v_\mu^{(c_k)}(t_k))^{-1}(\mathbf{e} - \mathbf{f})^T(\mathbf{v}^{(c_k)}(t_k) - \mathbf{v}^{(i)}(t_k))\right] \quad (15.27)$$

which is directly computable, since all terms of the r.h.s. are known.

After computing $\{\delta_k^\mu\}$ for all allocation cases of a traffic sample Θ, the empirical distribution of the results can be used to estimate the cumulative probability distribution of δ^μ. If we define $\delta^\mu(\varepsilon)$ as the inverse of the probability distribution, we can approximate it from the finite-sample estimate

$$\delta^\mu(\varepsilon) = \left(\delta^\mu \mid \mathrm{Prob}\left(c(\mathbf{u};\mathbf{e}^T) \neq c(\mathbf{u};\mathbf{e}^T(\delta^\mu))\right) \leq \varepsilon\right) \quad (15.28)$$

for a given ε. This function can be regarded as the range of perturbation that causes a given fraction of decisions to change from the original c_k to something else.

An example is shown graphically in Fig. 15.7: the $\delta^\mu(\varepsilon = 0.1)$ values are the intersections of the cumulative δ curves with the $\varepsilon = 0.1$ line. By a similar procedure, we can also calculate separate bounds for negative and positive perturbations (by modifying Equation (15.27)), and use the absolute minimum.

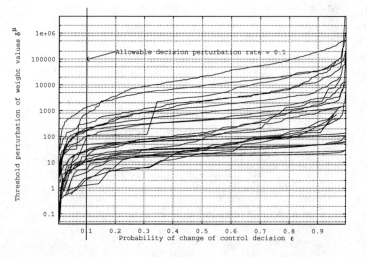

Fig. 15.7. Sensitivity of the control decisions w.r.t. the perturbation of the controller connection weights (example). The curves show the perturbation δ^μ needed to change the decisions with probability ε, as the function of ε, for 25 connection weights ($\mu = 1, \ldots, 25$).

The level of δ^μ that corresponds to $\varepsilon = 0.1$ determines the allowed range of perturbation for each connection weight coefficient e_μ in the output layer; as seen in Fig. 15.6, with 10% of changed (random) decisions, we can expect that if the performance deteriorates, it will be worse only on the order of about 10%. Fluctuation of the performance at this level is considered permissible for online, onsite adaptation.

15.4.3 Simulation Results
for Adaptive Optimal Group Control

The proposed method has been tested by several computer simulations. The elevator system was modeled by a simulation program, that took into account the details of the elevator dynamics (timings and logic of drive and door control, passenger movements etc). We have used a version of the "FLEX-8820 Fuzzy/AI Group Control System" of Fujitec Co., Ltd., as the teacher group control algorithm. The performance index H^0 obtained by this algorithm was also used as a reference, to evaluate the performance H of the proposed new method.

It has been found that prelearning of the waiting time prediction task Γ could succeed in faster adaptation [182], and then in all cases reported below, the adaptation was started with a pretrained network.

The optimal number of hidden units per elevator car q was estimated by testing the prediction error with different q. Although a higher q gave slightly better results, the increase after $q = r$ was not significant, hence this value ($q = r = 25$) was used. For pretraining, 2000 waiting time and allocation samples have been collected with the prototype controller. For adaptation, $n \in [100, 300]$ (hall calls/adaptation step), and $n_{\text{iter}} \in [200, 500]$ was used (for other parameters, see Section 15.4.1).

Simulation conditions for the examples below are given in Table 15.2. The

Table 15.2. Conditions for the test of the adaptation method

	Case 1	Case 2
Number of floors	15	15
Number of cars	6	6
Rated capacity [passg./car]	24	24
Rated speed [m/min]	150	150
Passenger O/D distribution	(see Table 15.3)	uniform
Arrival rate [passg./h]	600 - 1500	1500
Inter-arrival distribution	Poisson	Poisson

O/D distribution of Case 1 is reasonably typical, obtained from estimating the traffic flow from the recording of the elevator traffic data in a typical office building (see Table 15.3). Since the prototype controller was designed mainly for such typical traffic, it can provide very efficient control, and we cannot expect large improvements from the adaptive neurocontroller over the performance level H^0. On the other hand, Case 2 is an example of an untypical

Table 15.3. O/D distribution for Case 1. Numerals are relative values of traffic probabilities; "." means "0".

		Destination floors														
		B1	1	2	3	4	5	6	7	8	9	10	11	12	13	14
Origin floors	B1	.	1	.	1	.	1	.	.	1	.	.	.	.	.	.
	1	3	.	.	8	8	25	.	12	12	16	17	7	12	6	.
	2	.	.	.	.	.	.	.	.	.	.	.	.	2	.	.
	3	3	3	.	.	.	2	.	.	.	.	1	.	.	1	.
	4	3	3	3	.	.	.	3	.	.	1	3	6	3	.	.
	5	.	17	5	.	.	.	.	3	3	10	.	.	3	3	.
	6	.	4	.	.	2	.	.	2	.	2	.	.	.	.	.
	7	.	10	.	.	2	.	3	.	1	.	1	3	1	9	3
	8	.	5	.	.	.	5	.	.	.	5	2	.	.	2	.
	9	2	13	.	.	.	3	7	.	3	.	.	1	1	1	.
	10	2	8	2	.	.	2	.	4	2	2	.	.	.	3	.
	11	.	8	.	.	10	1	.	1	1	1	.	.	.	.	.
	12	.	7	1	.	2	4	1	5	2	.	1	.	.	2	.
	13	.	9	1	1	1	2	1	.	1	.	1	.	.	.	.
	14	.	.	.	.	.	.	.	.	.	.	.	.	.	.	.

(artificial) O/D distribution: all probability values are equal (we omit showing the obvious O/D table). We cannot expect the prototype controller to have control rules designed for such untypical traffic situations, so the possibility of improvement by the adaptive neurocontroller is higher.

Performance results for Case 1 are shown in Figs. 15.8 and 15.9 for two neurocontrollers. The first one was adapted to medium-level traffic conditions ($\lambda = 900$ passg./h), while the second one to more heavy traffic ($\lambda = 1500$ passg./h). The difference $\Delta H_i \equiv H_i - H_i^0$ of the performance values (average waiting time) of the new and the conventional controllers is plotted against the reference value H_i^0. Here, i is the index of test runs with different traffic samples Θ^i, but with the same average passenger arrival rate. Values for individual test runs $\Delta H_i(\lambda)$ are shown by open marks, and average values for all runs with the same traffic density $\langle \Delta H(\lambda) \rangle$ by solid marks.

Even in this case, both nets have somewhat better performance than the original "teacher" controller, even for some traffic conditions that are not encountered during adaptation. This indicates that the reinforcement learning has succeeded in generalizing from the training conditions to more general ones, except for low values of λ. The improvement of performance ($H \neq H^0$) has been found significant at levels between 95% and 99% by the χ^2 test, for $\lambda \in \{900, 1200, 1500\}$ [passg./h].

In Case 2, the adaptation process is shown in Fig. 15.10, and the performance improvement is shown in Fig. 15.11.

Performance results are summarized in Table 15.4. We can see that the performance improvement is large for such an untypical case. This was confirmed by other tests, too.

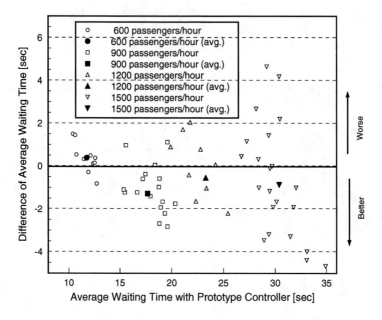

Fig. 15.8. Performance of the adapted controller for Case 1. The controller is trained with medium traffic conditions (900 passg./h).

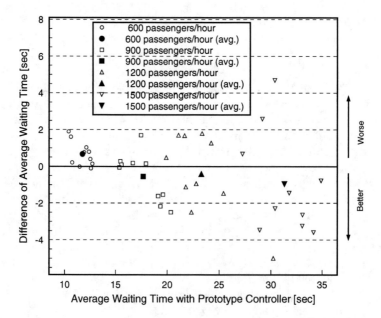

Fig. 15.9. Performance of the adapted controller for Case 1. The controller is trained with heavy traffic conditions (1500 passg./h).

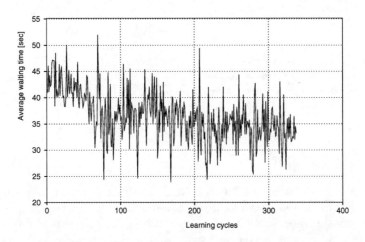

Fig. 15.10. Adaptation for Case 2. The controller is initialized with pre-training for Case 1. Performance levels (average waiting times for 100 hall calls) are tested at each training step.

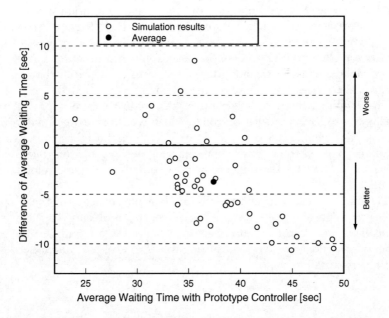

Fig. 15.11. Performance of the adapted controller for Case 2. The controller is adapted to the special traffic conditions of this case.

Table 15.4. Performance levels of the neurocontroller for Case 2

	Waiting time [s] avg. (std.dev.)
Prototype controller	37.41 (5.09)
Pretrained neurocontroller	37.88 (3.82)
Adapted neurocontroller (340 steps)	33.68 (4.02)

(46 traffic samples; 300 calls per sample)

15.5 Conclusions

We have seen that it is possible to build an elevator group controller by using neural networks, and it can be made adaptive by the reinforcement learning with stochastic approximation. The simulation results show that the adaptive controller can achieve better performance than a state-of-the-art commercial elevator group controller, especially for untypical traffic patterns.

Although we have determined the conditions for online adaptation, and online adaptation was found possible, the convergence rate is not yet satisfactory. More research is needed to improve the online adaptation speed. We also need to investigate online adaptation in the case of changing traffic conditions. The training schedule (a_ν, s_ν) is also not yet optimized, and it is not clear yet whether a restart process or a continuously running procedure is better. It is also possible to consider the adoption of the "score function method" [187] to this problem, as the next phase of this research.

The proposed method is general in the sense that we approximate an existing controller by a neurocontroller, which is then tuned by reinforcement learning. This simple scheme is made practically useful by using prelearning for improvement of the internal representation in the neurocontroller, and by sensitivity analysis for the range of perturbation during online reinforcement learning.

Therefore we can expect that the fundamental concept of the proposed method is applicable to a wider class of control problems. Possible future applications include the control of similar traffic systems, and also the application to other stochastic optimization problems, like, *e.g.*, online scheduling or online routing.

16

Optimal Control by Evolution Strategies and Particle Swarm Optimization

This chapter is devoted to the question: *How to use optimization algorithms in an effective and efficient manner?* Modern search heuristics have proved to be very useful for solving complex realworld optimization problems that cannot be tackled through classical optimization techniques [188]. Many of these search heuristics involve a set of exogenous parameters that affect their convergence properties. The determination of an adequate population size for evolution strategies is crucial for many optimization problems: Increasing the population size from 10 to 50 might improve the algorithm's performance—whereas a further increase might result in a performance decrease.

An optimal parameter setting, or statistically speaking, an optimal *algorithm design*, depends on the problem at hand as well as on the restrictions posed by the environment *i.e.* time and hardware constraints). Algorithm designs are usually either determined empirically or set equal to widely used default values.

We propose a methodology for the experimental analysis of optimization algorithms to determine improved algorithm designs and to learn how the algorithm works. The proposed technique employs computational statistic methods to investigate the interactions among optimization problems, algorithms, and environments.

Our focus lies on evolution strategies and particle swarm optimization. Both belong to the class of stochastic, population-based optimization algorithms [137]. They were introduced in Chapter 12.

An optimization practitioner is interested in robust solutions,*i.e.* solutions independent from the random seeds that are used to generate the random numbers during the optimization run. The proposed statistical methodology provides guidelines to design robust algorithms under restrictions, such as a limited number of function evaluations and processing units. These restrictions can be modeled by considering the performance of the algorithm in terms of the (expected) best function value for a limited number of function evaluations. A discussion on different problem classes for real-world optimization problems is provided in [189].

To justify the usefulness of our approach, we analyze the properties of several algorithms from the viewpoint of an optimization practitioner in the context of realworld optimization problems (optimization-via-simulation). More specifically, we consider the optimization of an elevator group controller as well as well-known test functions.

The S-ring, which was introduced in Chapter 14, can be used as a test problem generator for benchmark testing of algorithms, because it has only a few parameters. Furthermore, the S-ring also constitutes a well-suited model to simulate and analyze bunching problems [5]. We have compared the performance of the Kiefer-Wolfowitz stochastic approximation, the Q-learning algorithm, and the $(1 + 1)$-ES on the S-ring in Chapter 14. In the following, we will extend this comparison by including population-based search heuristics in our analysis.

16.1 Sequential Parameter Optimization

16.1.1 SPO as a Learning Tool

Theoretical results as well as heuristics to cope with noise have been presented in Chapter 12. But how can the optimization practitioner determine if these concepts work in practice? Hence, experiments are necessary. Experiment has a long tradition in science. To analyze experimental data, statistical methods can be applied. *Sequential parameter optimization* (SPO) was developed to tackle these problems in an effective and efficient manner [190]. SPO combines methods from classical *design of experiments* (DOE) and modern *design and analysis of computer experiments* (DACE).

It is not a trivial task to answer the final question "Is algorithm A better than algorithm B?" Results that are statistically significant are not automatically scientifically meaningful.

Example 16.1 (Floor and ceiling effects). The statistically meaningful result "all algorithms perform equally" can be scientifically meaningless, because the problem instances are too hard for any algorithm. A similar effect occurs if the problem instances are too easy. The resulting effects are known as floor or ceiling effects, respectively.   □

SPO provides numerical and graphical tools to test if the statistical results are really relevant or have been caused by the experimental setup only. It is based on a framework that permits a delinearization of the complex steps from raw data to scientific hypotheses. Substantive scientific questions are broken down into several local hypotheses that can be tested experimentally. The optimization process can be regarded as a process that enables learning. SPO consists of the twelve steps that are reported in Table 16.1. These steps and the necessary statistical techniques will be presented in the following. SPO has been applied on search heuristics in the following domains:

1. machine engineering: design of mold-temperature control [191–193]
2. aerospace industry: airfoil design optimization [194]
3. simulation and optimization: elevator group control [195]
4. technical thermodynamics: nonsharp separation sequences [196]
5. economy: agrienvironmental policy switchings [197]

Other fields of application are in fundamental research:

1. algorithm engineering: graph drawing [198]
2. statistics: selection under uncertainty (optimal computational budget allocation) for PSO [154]
3. evolution strategies: threshold selection und step-size adaptation [199]
4. computational intelligence: algorithmic chemistry [196]
5. particle swarm optimization: analysis und application [165]
6. numerics: comparison and analysis of classical and modern optimization algorithms [200]

Further projects, *e.g.* vehicle routing and door-assignment problems and the application of methods from computational intelligence to problems from bioinformatics are the subject of current research.

Table 16.1. Sequential parameter optimization (SPO). This approach combines methods from computational statistics and exploratory data analysis to improve (tune) the performance of direct search algorithms.

Step action
(S-1) Pre-experimental planning
(S-2) Scientific claim
(S-3) Statistical hypothesis
(S-4) Specification of the
(a) optimization problem
(b) constraints
(c) initialization method
(d) termination method
(e) algorithm (important factors)
(f) initial experimental design
(g) performance measure
(S-5) Experimentation
(S-6) Statistical modeling of data and prediction
(S-7) Evaluation and visualization
(S-8) Optimization
(S-9) Termination: If the obtained solution is good enough, or the maximum number of iterations has been reached, go to step (S-11)
(S-10) Design update and go to step (S-5)
(S-11) Rejection/acceptance of the statistical hypothesis
(S-12) Objective interpretation of the results from step (S-11)

Note that optimization practitioners do not always choose the absolute best algorithm. Sometimes a robust algorithm or an algorithm that provides insight into the structure of the optimization problem is preferred.

16.1.2 Tuning

In order to find an optimal algorithm design, or to tune the algorithm, it is necessary to define a performance measure. Effectivity (robustness) and efficiency can guide the choice of an adequate performance measure.

Optimization runs will be treated as experiments. From the viewpoint of an experimenter, design variables (factors) are the parameters that can be changed during an experiment. Generally, there are two different types of factors that influence the behavior of an optimization algorithm:

1. problem-specific factors, *e.g.*, the objective function.
2. algorithm-specific factors, *e.g.*, the population size or other exogenous parameters.

We will consider experimental designs that comprise problem-specific factors and exogenous algorithm-specific factors. Algorithm-specific factors will be considered first. *Endogenous* can be distinguished from *exogenous parameters* [139]. The former are kept constant during the optimization run, whereas the latter, *e.g.* standard deviations, are modified by the algorithms during the run. Consider $\mathcal{D}_A$, the set of all parameterizations for one algorithm. An *algorithm design* X_A is a set of vectors, each representing one specific setting of the design variables of an algorithm. A design can be specified by defining ranges of values for the design variables. Note that a design can contain none, one, several or even infinitely many design points. The *optimal algorithm design* is denoted as X_A^*. The term "optimal design" can refer to the best design point x_a^* as well as the most informative design points [142, 201].

Let $\mathcal{D}_P$ denote the set of all problem instances for one optimization problem. *Problem designs* X_P provide information related to the optimization problem, such as the available resources (number of function evaluations) or the problem's dimension.

An *experimental design* $X_e \in \mathcal{D}$ consists of a problem design X_P and an algorithm design X_A. The run of a stochastic search algorithm can be treated as an experiment with a stochastic output $Y(x_a, x_p)$, with $x_a \in \mathcal{D}_A$ and $x_p \in \mathcal{D}_P$. If the random seed is specified, the output would be deterministic. This case will not be considered further, because it is not common practice to specify the seed that is used in an optimization run. Performance can be measured in many ways, for example as the best or the average function value for n runs.

Goals of the experimental approach from this chapter can be stated as follows:

(G-1) *Efficiency.* To find a design point $x_a^* \in \mathcal{D}_A$ that improves the performance of an optimization algorithm for one specific problem design point $x_p \in \mathcal{D}_P$.

(G-2) *Robustness.* To find a design point $x_a^* \in \mathcal{D}_A$ that improves the performance of an optimization algorithm for several problem design points $x_p \in \mathcal{D}_P$.

Statistical techniques to attain these goals will be presented next. Classical regression techniques will be considered first.

16.1.3 Stochastic Process Models as Extensions of Classical Regression Models

The classical DOE approach consists of three steps: Screening, modeling, and optimization. Each step requires different experimental designs. Linear regression models are central elements of the classical design of experiments approach [202, 203]. We will discuss a different approach that relies on a stochastic process model: each algorithm point with associated output is treated as a realization of a stochastic process. *Kriging* is an interpolation method to predict unknown values of a stochastic process and can be applied to interpolate observations from computationally expensive simulations. Our presentation follows concepts introduced in [204–206].

Consider a set of m design points $x = (x^{(1)}, \ldots, x^{(m)})^T$ with $x^{(i)} \in \mathbb{R}^d$. In the *design and analysis of computer experiments* (DACE) *stochastic process model,* a deterministic function is evaluated at the m design points x. The vector of the m responses is denoted as $y = (y^{(1)}, \ldots, y^{(m)})^T$ with $y^{(i)} \in \mathbb{R}$. The process model proposed in [204] expresses the deterministic response $y(x^{(i)})$ for a d-dimensional input $x^{(i)}$ as a realization of a regression model $\mathcal{F}$ and a stochastic process Z,

$$Y(x) = \mathcal{F}(\beta, x) + Z(x) \tag{16.1}$$

DACE Regression Models

We use q functions $f_j : \mathbb{R}^d \to \mathbb{R}$ to define the regression model

$$\mathcal{F}(\beta, x) = \sum_{j=1}^{q} \beta_j f_j(x) = f(x)^T \beta \tag{16.2}$$

Regression models with polynomials of orders 0, 1, and 2 have been used in our experiments. The constant regression model with $q = 1$ reads $f_1(x) = 1$, the linear model with $q = d + 1$ is $f_1(x) = 1, f_2(x) = x_1, \ldots, f_{d+1}(x) = x_d$, and the quadratic model: $f_1(x) = 1; f_2(x) = x_1, \ldots, f_{d+1}(x) = x_d; f_{d+2}(x) = x_1 x_1, \ldots, f_{2d+1}(x) = x_1 x_d; \ldots; f_q(x) = x_d x_d$. Regression models with a constant term only have been applied successfully to model the data and to predict new data points in the sequential approach.

DACE Correlation Models

The random process $Z(\cdot)$ (Equation 16.1) is assumed to have mean zero and covariance $V(w, x) = \sigma^2 \mathcal{R}(\theta, w, x)$ with process variance σ^2 and correlation model $\mathcal{R}(\theta, w, x)$. Correlations of the form

$$\mathcal{R}(\theta, w, x) = \prod_{j=1}^{d} \mathcal{R}_j(\theta, w_j - x_j)$$

will be used in our experiments. The correlation function should be chosen with respect to the underlying process [207]. Reference [208] discusses 7 different models. The Gaussian correlation function is a well-known example. It is defined as

$$\text{GAUSS}: \qquad \mathcal{R}_j(\theta, h_j) = \exp(-\theta_j h_j^2) \qquad (16.3)$$

with $h_j = w_j - x_j$, and for $\theta_j > 0$.

The regression matrix R is the matrix with elements

$$R_{ij} = R(x_i, x_j) \qquad (16.4)$$

that represent the correlations between $Z(x_i)$ and $Z(x_j)$. The vector with correlations between $Z(x_i)$ and a new design point $Z(x)$ is

$$r(x) = (R(x_1, x), \ldots, R(x_m, x)) \qquad (16.5)$$

Large θ_js indicate that variable j is active: function values at points in the vicinity of a point are correlated with Y at that point, whereas small θ_js indicate that distant data points also influence the prediction at that point. The *empirical best unbiased linear predictor* (EBLUP) can be shown to be

$$\hat{y}(x) = f^T(x)\hat{\beta} + r^T(x)R^{-1}(y - F\hat{\beta}) \qquad (16.6)$$

where

$$\hat{\beta} = \left(F^T R^{-1} F \right)^{-1} F^T R^{-1} y \qquad (16.7)$$

is the generalized least-squares estimate of β in Equation (16.1), $f(x)$ are the q regression functions in Equation (16.2), and F represents the values of the regression functions in the m design points.

Maximum likelihood estimation methods to estimate the parameters θ_j of the correlation functions from Equation (16.3) are discussed in [208]. DACE methods provide an estimation of the prediction error on an untried point x, the *mean squared error* (MSE) of the predictor

$$\text{MSE}(x) = E\left(\hat{y}(x) - y(x) \right) \qquad (16.8)$$

The stochastic process model, which was introduced as an extension of the classical regression model, will be used in our experiments. Next, we have to decide how to generate design points, *i.e.* which parameter settings should be used to test the algorithm's performance.

Expected Improvement

Often, designs that use sequential sampling are more efficient than designs with fixed sample sizes. Therefore, we specify an initial design $X_A^{(0)} \in \mathcal{D}_A^{(0)}$ first. Information obtained in the first runs can be used for the determination of the second design $X_A^{(1)}$ in order to choose new design points more efficiently.

Sequential sampling approaches with adaptation have been proposed for DACE. For example, in [204] sequential sampling approaches with and without adaptation were classified to the existing metamodel. We will present a sequential approach that is based on the expected improvement. In [142, p. 178] a heuristic algorithm for unconstrained global minimization problems is presented. Consider one problem design X_P. Let $y_{\min}^k$ denote the smallest known minimum value after k runs of the algorithm, $y(x)$ be the algorithm's response, *i.e.* the realization of $Y(x)$ in Equation (16.1), and let x_a represent a specific design point from the algorithm design X_A. Then the improvement is defined as

$$
\text{improvement at } x_a = \begin{cases} y_{\min}^k - y(x_a), & y_{\min}^k - y(x_a) > 0 \\ 0, & \text{otherwise} \end{cases} \tag{16.9}
$$

for $x_a \in \mathcal{D}_A$.

As $Y(\cdot)$ is a random variable, its exact value is unknown. The goal is to optimize its expectation, the so-called *expected improvement*. New design points, which are added sequentially to the existing design, are attractive "if either there is a high probability that their predicted output is below [minimization] the current observed minimum and/or there is a large uncertainty in the predicted output." This leads to the *expected improvement heuristic* shown in Algorithm 16.1.

Algorithm 16.1 Expected improvement heuristic

1: Set $t = 0$. Choose an initial design $X_A^{(t)} \in \mathcal{D}_A$ with n points.
2: Run the algorithm at $x_i \in X_A^{(t)}$, $i = 1, \ldots, n$, to obtain the vector of output values $y(x)$.
3: Check the termination criterion.
4: Select a new point x_{n+1} that maximizes the expected improvement.
5: Run the algorithm at x_{n+1} to obtain the output $y(x_{n+1})$.
6: Set $X_A^{(t+1)} = X_A^{(t)} \cup \{x_{n+1}\}$, $n = n + 1$, and go to 3:

Sensitivity Analysis

The stochastic process model from Equation 16.1 can be used to estimate the effect of certain factors. Reference [142] recommends to use a small design to determine important factor levels. After running the optimization algorithm, scatterplots of each input versus the output can be analyzed. Reference [209] advocates the use of sensitivity analysis. A screening algorithm that is similar in spirit to forward selection in classical regression analysis is used to identify important factors. Reference [204] proposes an ANOVA-type decomposition of the response into an average, main effects for each factor, two-factor interactions and higher-order interactions. Let the average of $y(x)$ over the experimental region be

$$\mu_0 = \int y(x) \prod_h dx_h$$

define the main effect of factor x_i averaged over the other factors by

$$\mu_i(x_i) = \int y(x) \prod_{h \neq i} dx_h - \mu_0$$

and the interaction effect of x_i and x_j by

$$\mu_{ij}(x_i, x_j) = \int y(x) \prod_{h \neq i,j} dx_h - \mu_0 - \mu_i(x_i) - \mu_j(x_j)$$

To illustrate these equations, the reader may think of the evolution strategy introduced in Table 12.3. It has has nine exogenous parameters (design variables). The variables $x_1, x_2, \ldots, x_9$ can be used for their representation. The effects show which factors, *e.g.* population size or standard deviations, have the greatest impact on the performance. Higher-order interactions can be obtained accordingly. To estimate these effects, $y(x)$ is replaced by $\hat{y}(x)$. Factors for which the predicted response is not sensitive to, can be fixed in subsequent modeling steps.

In the following, we used a similar approach that was proposed by [210] to plot the estimated effects of a subset x_{effect} of the x variables. The effect of a factor or a set of factors is defined as

$$\mu(x_{\text{effect}}) = \frac{1}{V} \int Y(x) dx_{\text{out}}$$

where V is the volume of the x_{out} region over which is integrated. Note that $Y(x)$ has to be replaced by $\hat{Y}(x)$ from Equation (16.6).

The effect $\mu(x_{\text{effect}})$ can be determined numerically as

$$\mu(x_{\text{effect}}) = \frac{1}{n} \sum_{i=1}^{n} Y\left(x_{\text{effect}}, x_{\text{out}}^{(i)}\right) \tag{16.10}$$

where n denotes the number of points $x_{\text{out}}^{(1)}, \ldots, x_{\text{out}}^{(n)}$ of a grid representing the d-dimensional x_{out} space[1].

---

[1] Equation (16.10) is Equation (3.5) in [210].

Based on theorems from [210, p. 22] we implemented a program to estimate and plot the main factor effects. Furthermore, three-dimensional visualizations produced with the DACE toolbox [206] can be used to illustrate the interaction between two design variables and the associated mean squared error of the predictor.

16.1.4 Space-filling Designs

Latin hypercube sampling was used to generate the initial algorithm designs. Consider n levels being examined and k design variables. A Latin hypercube is a matrix of n rows and k columns. The k columns contain the levels 1, 2, ..., n, randomly permuted, and the k columns are matched at random to form the Latin hypercube. The resulting *Latin hypercube designs* are space-filling designs, as can be seen in Fig. 16.1. Reference [211] introduced LHDs for computer experiments, while Ref. [142] gives a comprehensive overview.

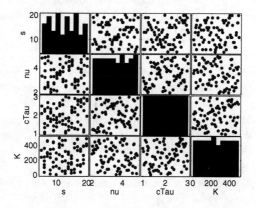

Fig. 16.1. Space filling design. Eighty design points.

16.2 The S-ring Model as a Test Generator

The S-ring model can be used to generate test problem instances. An S-ring problem instance can be characterized by the number of sites, the number of elevator cars, the arrival probability, and the simulation time, see Table 14.1 and Table 16.2. Note, that we are using the S-ring-model to define a minimization problem: the (expected) number of sites with waiting customers

$$y = \sum_{i=0}^{n-1} c_i \tag{16.11}$$

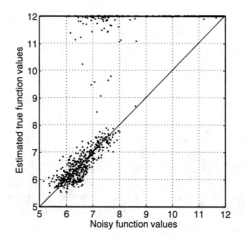

Fig. 16.2. S-ring. Estimated versus noisy function values. Test instance $x^{(1)}_{(12,4,0.2)\text{sring}}$ as listed in Table 16.2. Estimated values have been gained through re-evaluation, whereas noisy function values are based on one evaluation only. Points representing values from functions without noise would lie on the bisector.

In Equation 14.1, the S-ring-model was used to define an objective function h for a maximization problem. Both optimization problems are equivalent, because $h = n - y$. Therefore, the values from Table 14.2 can be used for the $(6,2,0.3)$-sring function in this chapter. Figure 16.2 illustrates the influence of noise in the S-ring model. For an optimizer small vales from the range -5 to 5 have been added to test the robustness of the solutions. As can be seen from Fig. 16.3, the solution is relatively robust, even for different problem instances. However, if randomly generated values are added to the optimizer, the smoothness disappears.

A problem design as introduced in Section 16.1.2 specifies one or more instances of an optimization problem and related restrictions, $e.g.$ the number of

Table 16.2. Test instances for the S-ring model. Note that the (n, m, p)-sring has been introduced on p. 158 in Chapter 14. The simulation time refers to a simulation run of N steps as described on p. 160.

Instance	Dimension	Number of sites: n	Number of elevator cars: m	Arrival probability: p	Simulation time: N
$(6,2,0.3)$-sring	12	6	2	0.3	1000
$(12,4,0.2)$-sring	24	12	4	0.2	1000
$(18,8,0.2)$-sring	36	18	8	0.2	1000
$(24,16,0.3)$-sring	48	24	16	0.3	1000
$(48,32,0.3)$-sring	96	48	32	0.3	1000

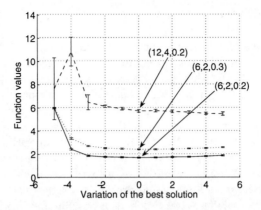

Fig. 16.3. S-ring. Robustness of solutions for the S-ring-problem instances. The best solution determined by algorithm design x^*_{ES2} from Table 16.5 was varied by adding $i \times 1$, $i = -5, -4, -3, \ldots, 5$. The y value associated to the value 0 on the abscissa represents the function value of the best solution x, that has not been disturbed. That is, values on the x-axis denote the variation strength. If small, but randomly generated values are added to x, the smoothness of the curves is destroyed.

available resources (function evaluations). The problem design from Table 16.3 has been used for the first experimental studies. To keep the experimental setup realistic, we have chosen a budget of 1000 function evaluations only. Note, that different settings, especially $t_{max} \gg 1000$, might produce directly opposed results. However, in many real-world applications, more than 1000 function evaluations are prohibitive. The deterministic initialization scheme DETEQ, which uses one prespecified starting point $x^{(0)}$, has been selected to reduce the variance and to enable the application of ANOVA techniques. The algorithm terminates, if its budget (the maximum number of function evaluations, *e.g.* $t_{max} = 1000$) is exhausted. The mean best function value (MBST) has been chosen as a performance measure, because it is commonly used.

Table 16.3. Problem design for the S-ring experiments. This problem design has been used as the first design to determine suitable algorithm designs.

Design	n	t_{max}	d	Init.	Term.	$x^{(0)}$	PM
$x^{(1)}_{(6,2,0.3)\text{sring}}$	50	1000	12	DETEQ	EXH	100	MBST

16.3 Experimental Results for the S-ring Model

At each stage, Latin hypercube designs are used. Reference [212] reports that experience with the stochastic process model had indicated that 10 times the expected number of algorithm design variables is often an adequate number of runs for the initial LHD. That is, for the particle swarm optimization with constriction coefficient, which requires four design variables (see Section 12.3.1), 40 runs are recommended. Reference [167] described a final comparison of two PSO variants, a *Nelder–Mead simplex* (NMS) algorithm, and a quasi-Newton method on the (6,2,0.3)-sring problem instance. Both PSO variants outperformed the other algorithms. Here, we will extend this analysis including further algorithms and additional problem instances to our experiments. The performance of the $(1 + 1)$-ES defines a baseline for these comparisons.

One goal of our experiments is to find an answer to the question: *Does one algorithm perform best on several instances of the S-ring?*

16.3.1 Evolution Strategies

The $(1 + 1)$-ES

The $(1 + 1)$-ES was not able to cope with the S-ring function. In any setting, the step sizes decreased too rapidly. The best found function value reads 2.61, which is no improvement compared with the function value of the initial configuration.

The $(1 + 1)$-ES with Threshold Acceptance

Table 16.4 shows the algorithm design for the $(1+1)$-ES with threshold acceptance. x_{TA}^* denotes the improved design that has been determined with SPO. Threshold acceptance prevents small step sizes, so that the search can be continued, even if no real progress occurs. Introducing the threshold acceptance method improves the algorithm performance on this problem instance compared to the simple $(1+1)$-ES. The interaction between threshold acceptance and step-size adaptation is shown in Fig. 16.4.

Table 16.4. Algorithm designs for the evolution strategy with threshold adaptation. $x_{TA}^{(l)}$ and $x_{TA}^{(u)}$ denote the lower and upper bounds to generate the LHD, respectively, and x_{TA}^* denotes the parameter settings of the improved design that was found by the sequential approach.

Design	s_n	s_r	s_a	s_u
$x_{TA}^{(l)}$	5	3	0.6	2
$x_{TA}^{(u)}$	30	8	0.85	20
x_{TA}^*	18	5.96	0.62	5

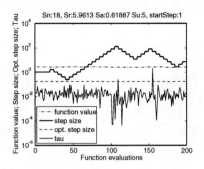

Fig. 16.4. S-ring. $(1 + 1)$-ES. Threshold acceptance prevents too small step sizes. The progress in the function values is relatively small compared to the fluctuations in the threshold values and standard deviations. The optimal step size has been determined for a similar situation on the sphere function. The step size, which has been used by the $(1 + 1)$-TA, is larger than the optimal step size—a phenomenon that could be observed for many optimization scenarios.

The experiments reveal that TA prevents too small steps size. However, Fig. 16.4 illustrates that TA was not able to find better solutions. Population-based algorithms will be considered next, because the individual-based algorithms showed a relatively poor performance on the S-ring problem.

Multimembered Evolution Strategies

Sequential parameter optimization was used to tune the algorithm designs. The problem design from Table 16.3 and the algorithm designs from Table 16.5 have been used to set up the experimental designs. This is step (S-4) in Fig. 16.1. A LHD with 80 design points has been chosen as the initial design. Each design point has been evaluated 5 times. Additional design points—that maximize the expected improvement—have been added sequentially. The best algorithm design after 500 algorithm runs is reported in Table 16.5. Six ES variants were compared:

1. ES with one step size and without recombination: x_{ES1}
2. ES with one step size and local recombination: x_{ES2}
3. ES with one step size and global recombination: x_{ES3},
4. ES with d step sizes and without recombination: x_{ES4},
5. ES with d step sizes and local recombination: x_{ES5}, and
6. ES with d step sizes and global recombination: x_{ES6}.

Table 16.5 summarizes the improved algorithm designs for these six algorithms and Table 16.6 compares statistics based on 50 runs.

Experimental Analysis

The results from Table 16.6 indicate that algorithm design x_{ES3}^*, *i.e.* ES with one step size and global recombination, performs best on problem design

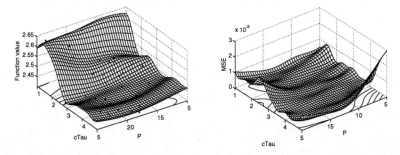

Fig. 16.5. S-ring. Evolution strategy with one step-size. Local intermediate recombination. Left: Interaction between population size and learning rate τ. Dots denote sample points. Right: mean squared prediction error. The MSE is high in regions where no points have been sampled.

Table 16.5. Algorithm designs for the evolution strategy. These algorithm designs have been tuned on the 12-dimensional (6,2,0.3)-`sring` problem instance. The best algorithm design is printed in *boldface*.

Algorithm-design	σ_n	**reco**	s	ν	c_τ	κ
$x_{\mathrm{ES}}^{(l)}$			5	2	1	1
$x_{\mathrm{ES}}^{(u)}$			20	5	5	100
x_{ES1}^{*}	1	no	15	4.19	3.90	52
x_{ES2}^{*}	1	local	13	4.69	2.03	2
x_{ES3}^{*}	**1**	**global**	**15**	**4.35**	**2.39**	**2**
x_{ES4}^{*}	d	no	10	3.37	3.74	43
x_{ES5}^{*}	d	local	6	4.84	3.63	21
x_{ES6}^{*}	d	global	15	4.30	2.13	13

Table 16.6. Results for the algorithm designs for the evolution strategy optimizing the 12-dimensional (6,2,0.3)-`sring`. Each run was repeated 50 times. The best results are printed in *boldface*. Algorithm design x_{ES3}^{*} performs best.

Algorithm-design	Mean	Median	SD	Min	Max	Minboot
x_{ES1}^{*}	2.4577	2.4538	0.0256	2.4179	2.5198	2.4375
x_{ES2}^{*}	2.4124	2.4116	0.0115	2.3902	**2.4421**	2.4030
x_{ES3}^{*}	**2.4095**	**2.4084**	**0.0103**	**2.3877**	2.4422	**2.4017**
x_{ES4}^{*}	2.4351	2.4334	0.0135	2.4083	2.4737	2.4249
x_{ES5}^{*}	2.4316	2.4327	0.0128	2.4011	2.4721	2.4213
x_{ES6}^{*}	2.4335	2.4298	0.0248	2.4035	2.5734	2.4202

$x_{(6,2,0.3)\text{sring}}^{(1)}$ from Table 16.3). In the following we will present some statistical tools that can be useful to answer the question why this algorithm design performs best.

Effect plots illustrate that larger learn rates improve the performance. This can also be seen from the correlation function parameters θ in the stochastic process model (see Equation (16.1)): $\theta_1 = 0.3337$, $\theta_2 = 0.1530$, $\theta_3 = 3.1748$, and $\theta_4 = 0.2042$, the generalized least-squares estimate $\beta = 0.2427$, and the estimate of the process variance is $\sigma = 0.0034$. Effect plots for the remaining five algorithms show a similar behavior.

A plot displaying the interaction between learning rate and population size is shown in Figure 16.5. It supports the assumption that the learning rate has a stronger influence than the population size.

Trellis plots depict relationships between different factors through conditioning. They show how plots of two factors change with variations in a third, the so called conditioning factor. Trellis plots consist of a series of panels where each panel represents a subset of the complete data divided into subintervals of the conditioning variable. Three intervals for c_τ in Fig. 16.6 have been used: $I_1 = [1.009, 2.155]$, $I_2 = [2.1539, 3.3421]$, and $I_3 = [3.3408, 4.9988]$. Each interval contains 170 data points. There is an overlap of 5 data points between adjacent intervals, because 500 data points are distributed among three intervals.

Figure 16.7 visualizes the influence of a different number of step sizes and recombination schemes. This comparison is based on the tuned algorithm designs from Table 16.5. These histograms indicate that d step sizes result in a worse performance of the algorithms and that local recombination is advantageous.

The analysis reveals that one step size might be advantageous. A possible explanation for this phenomenon is the relatively small number of function evaluations that prevent a step-size adaptation for several step sizes.

Recombination

Figure 16.8 compares the best object variables x found by an ES with and an ES without intermediate recombination. These x_i values have been normalized, that is $0 \leq x_i \leq +1$, because only their relative values are of importance. This can be seen from Equation (14.5)

$$\pi(x) = \theta(w^T \cdot x) \tag{16.12}$$

so that $\pi(x) = \pi(c \cdot x)$ for any $c \in \mathbb{R}$.

Intermediate recombination prevents outliers, the diversity in the population is reduced. This effect improves the algorithm performance on the S-ring problem.

Compare scatter plots of the object variables representing the best position found during one run. Only the first four variables x_1 to x_4 are shown in

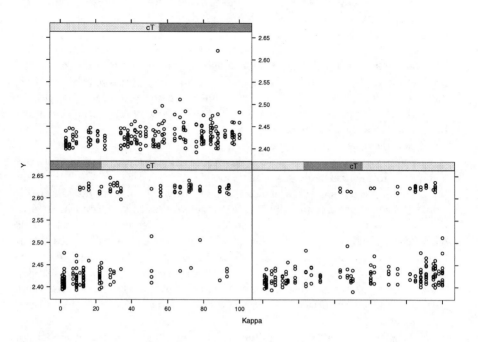

Fig. 16.6. S-ring. Evolution strategy. Trellis plots. Higher c_τ values improve the algorithm's performance. The data points have been divided into three intervals due to their c_τ values: $I_1 = [1.009, 2.155]$ with 170 data points, $I_2 = [2.1539, 3.3421]$ with 170 data points, and $I_3 = [3.3408, 4.9988]$ with 170 data points. These figures support the assumption that large c_τ values and small κ values improve the algorithm's performance.

Fig. 16.8 to improve readability. Squares denote object variables with associated function values that belong to the 0.1 quantile (these can be regarded as the best solutions). Each panel contains 500 data points. The interval I represents the space of possible object parameter values. These values have been normalized. Some coordinates require settings close to zero (in the center of the interval $I = [0, 1]$), e.g. x_1 and x_2, whereas others require settings at the border to generate very good solutions, e.g. x_3 and x_4. The *genetic repair hypothesis* [213] might give an explanation for this observation: "Not the different (desired) features of the different parents flow through the application of the recombination operator into the offspring, but their common features."

Discussion of the ES Results

Population-based algorithms with relatively high learning rates, one step size, and intermediate recombination performed best on the S-ring. Noise and large plateaus with equal function values reduced the step sizes in the $(1 + 1)$-

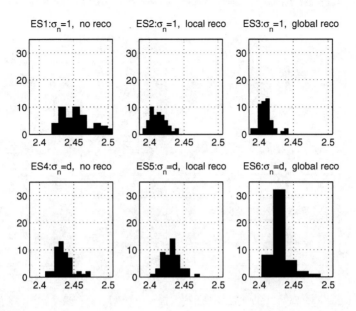

Fig. 16.7. S-ring. Comparison of the performance of different ES-settings. The 12-dimensional (6,2,0.3)-`sring` was used for these experiments. Evolution strategies with one step-size and global recombination (ES3) perform best.

ES. Threshold acceptance prevents this decrease, but the algorithm did not learn the correct step sizes, it performed a random walk. Population-based algorithms were able to combine useful information (Beyer's genetic repair effect).

The low number of function evaluations (1000) prevents a step-size adaptation for algorithms with twelve step sizes. They were outperformed by algorithms with one step size only.

16.3.2 Particle Swarm Optimization on the S-ring Model

Particle Swarm Optimization with OCBA

A variant of the particle swarm optimization algorithm that integrates the OCBA sequential selection method is presented next. We will denote this variant by PSO_{OCBA}. The variant PSO_{OCBA} is similar to the $PSO_{default}$ variant, except the determination of the swarm's global best and particles' personal best positions. In this part of the algorithm, the OCBA technique is applied. PSO_{OCBA} was introduced in [154] to compare its performance on the noisy sphere function $f(x) = \sum x_i^2$. In every iteration, there is a set of positions that are candidates for the swarm's global best. There are the new generated positions and the old personal-best positions of the particles. In the presence

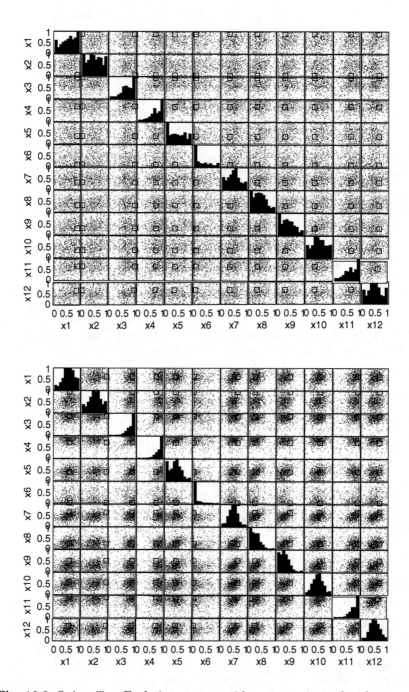

Fig. 16.8. S-ring. Top: Evolution strategy with one step size and without recombination. Bottom: Evolution strategy with one step size and global intermediate recombination. Scatter plot of the object variables x_1 to x_{12}. Squares denote positions with function values smaller than the lower 0.01 percent quantile. These plots illustrate the effect of recombination: it generates structure and extracts common features.

of noise, it is not obvious which of them is to choose as the swarm's global best. The OCBA approach has the aim to select the best within a set of possibilities with a high probability of correct selection under the restriction of a limited number of samples. It is used here to find the swarm's global best among the set of positions considered in the iteration.

With the design of the PSO$_{OCBA}$ algorithm, we aim at two objectives:

1. an increased probability to select the swarm's global best correctly
2. an increased probability to select the particles' personal bests correctly (as a byproduct of the repeated function evaluations of candidate positions by the OCBA method)

Table 16.7. Algorithm designs for the particle swarm optimization. These algorithm designs have been tuned on the 12-dimensional (6,2,0.3)-`sring` problem instance.

Algorithm-design	s	χ	φ	v_{max}
$x_{PSOC}^{(l)}$	3	0.8	4.0	5
$x_{PSOC}^{(u)}$	30	0.95	5.0	1000
x_{PSOC}^{def}	20	0.729	4.1	100
x_{PSOC}^{*}	15	0.884	4.354	103.668

Table 16.8. Results for the algorithm designs for the particle swarm optimization optimizing the 12-dimensional (6,2,0.3)-`sring`. Each run was repeated 50 times. Algorithm design x_{PSOC}^{*} performs best.

Algorithm-design	Mean	Median	Std	Min	Max	Minboot
x_{PSOC}^{def}	2.6218	2.6202	0.0151	2.5935	2.6507	2.6090
x_{PSOC}^{*}	2.4639	2.4441	0.0545	2.4081	2.6578	2.4308

The procedure is as follows. In the beginning all new positions are evaluated n_0 times, where n_0 is an exogenous parameter of the PSO$_{OCBA}$ variant. It is the number of repeated function evaluations for each position. The particles' personal bests have already received at least n_0 evaluations in one or more previous iterations. These at least n_0 evaluations, which for each position in consideration have been made, are used as an estimation for means and variances. In the next step, the OCBA method is used to assign an additional number of function evaluations to the positions. The additional evaluations are accomplished in rounds. There is a total number of additional function evaluations n_{add} and an incremental amount n_{inc}, with $1 \leq n_{inc} \leq n_{add}$. In each round n_{inc} evaluations are assigned to the candidate positions by the

OCBA method. The resulting function values are used to update the sample means and variances for the next round. The loop is exited after n_{add} additional function evaluations have been made.

In accordance with the OCBA technique, the position with the lowest mean of the function values is selected as the swarm's global best. The new personal bests of the particles result from the comparison between the function value means of their old personal best and new positions. All objective function evaluations made for the new personal-best position are stored for usage in the next generation. Algorithm 16.2 shows the algorithm PSO_{OCBA} as pseudocode. Reference [153] provides a code example of the OCBA method. Our implementation was inspired by it.

Algorithm 16.2 Pseudocode of the PSO_{OCBA} algorithm variant

1: Initialize swarm positions x and swarm velocities v
2: **repeat**
3:    Update particle velocities
4:    Update particle positions
5:    Draw initial n0 samples of new positions
6:    Update means and variances for new positions
7:    candidates = new positions and personal-best positions
8:    nAdd = Additional Samples
9:    **while** nAdd > 0 **do**
10:      Assign with OCBA nInc additional samples to candidates
11:      Draw additional samples
12:      Update means and variances
13:      nAdd = nAdd - nInc
14:    **end while**
15:    **if** new position better than personal-best position **then**
16:      remember new position as personal-best position
17:      global best position = best of particles' personal best
18:    **end if**
19: **until** termination criterion reached

Table 16.7 summarizes the default and improved particle swarm optimization algorithm designs and Table 16.8 compares statistics based on 50 runs. The experimental results show that a small number of additional function

Table 16.9. Additional exogenous parameters for the OCBA

Symbol	Parameter	Range	Default
n_0	initial number of function evaluations	$\mathbb{N}$	5
n_{add}	additional function evaluations each iteration, assigned using OCBA	$\mathbb{N}$	50
n_{inc}	incremental evaluations each OCBA round	$\mathbb{N}$	10

Table 16.10. Algorithm designs for the particle swarm optimization. These algorithm designs have been tuned on the 12-dimensional (6,2,0.3)-`sring` problem instance.

Algorithm design s		χ	φ	v_{max}	n_0	n_{add}	n_{inc}
$x_{PSOCOCBA}^{(l)}$	3	0.68	3.0	1	1	1	1
$x_{PSOCOCBA}^{(u)}$	30	0.9	6.0	750	5	20	5
$x_{PSOCOCBA}^{def}$	20	0.729	4.1	100	5	50	10
$x_{PSOCOCBA}^{*}$	14	0.883	5.953	639.892	1	11	3

Table 16.11. Results for the algorithm designs for the particle swarm optimization optimizing the 12-dimensional (6,2,0.3)-`sring`. Each run was repeated 50 times. Algorithm design $x_{PSOCOCBA}^{*}$ performs best.

Algorithm design	Mean	Median	Std	Min	Max	Minboot
$x_{PSOCOCBA}^{def}$	2.6248	2.6256	0.0157	2.5928	2.6593	2.6115
$x_{PSOCOCBA}^{*}$	2.4414	2.4367	0.0289	2.3949	2.5435	2.4221

evaluations, e.g., $n_0 = 1$, should be used to improve the algorithm's performance. Introducing OCBA alone is not sufficient.

Comparing the default, tuned, OCBA, and the tuned OCBA particle swarm optimizers (Tables 16.10 and 16.11) indicates that the combination of tuning and OCBA performs best. But is the performance of the PSO with OCBA ($x_{PSOCOCBA}^{*}$) really better than the performance of the tuned PSO (x_{PSOC}^{*})? The optimization practitioner has to decide whether this difference is meaningful. Regarding the objective interpretation of the results (this is step (S-12) in Table 16.1), we can use plots of the observed significance as shown in Figure 16.9. Consider $n = 500$: If the true difference in means is as large as $\delta = 0.005$ or larger, then the hypothesis $H = \delta$ is wrongly rejected 10% of the time. This observed p value increases, if only $n = 10$ or $n = 50$ repeated runs are performed. We can conclude that there is a difference in means, but its detection requires many experiments. Therefore it is doubtful if the difference in the performance of the two algorithms is meaningful. Further problems related to the comparison of algorithms are discussed in [190], while [214] presents an approach from statistics to tackle these questions.

Summary: Noise and PSO

We have examined the influence of noise on the performance of PSO, and compared several algorithmic variants with default and tuned parameters. Based on our results, we make the following conclusions:

1. Parameter tuning alone cannot eliminate the influence of noise.
2. Sequential selection procedures such as OCBA can significantly improve the performance of particle swarm optimization in noisy environments.

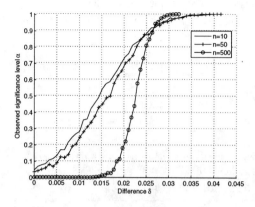

Fig. 16.9. Observed significance plots illustrate the relationship between p values, sample sizes, and differences in means for the tuned particle swarm optimization with and without OCBA

Local information plays an important role in this selection process and cannot be omitted.

Why did sequential selection methods improve the algorithm's performance? First, the selection of the swarm's best of one iteration was correct with a higher probability compared to re-evaluation approaches. Second, as more samples were drawn for promising positions, positions that remained and reached the next iteration were likely to have received more samples than the average. Samples accumulated and led to a greater sample base for each iteration's decisions. These two advantages might be transferable to other population-based search heuristics, in which individuals can survive several generations.

Summarizing, we can conclude that the differences between the tuned PSO and the PSO with OCBA is relatively small. The highest performance improvement was caused by the combination of SPO and OCBA, but the optimization practitioner has to decide whether this difference is of practical relevance.

16.4 Classical Algorithms on the S-ring Model

We summarize results from algorithm runs from classical algorithms on the S-ring. Reference [200] demonstrated that a Nelder–Mead simplex algorithm and a quasi-Newton (QN) method were not able to escape from local optima while optimizing the S-ring. In contrast to the artificial test functions from commonly used testsuites, real-world optimization problems often have many local minima on flat plateaus. The distribution of local optima in the search space is unstructured. Therefore these algorithms were unable to escape plateaus of equal function values. This behavior occurred independently of the parameterization of their exogenous strategy parameters.

Reference [200] noted: "Experimental results indicate that there is no generic algorithm that works equally well on each problem. Even different instances of one problem may require different algorithms, or at least different parameterizations of the employed algorithms. None of the algorithms has proved in our study to be satisfying for every problem. The quasi-Newton method, as expected, outperformed the other algorithms on the Rosenbrock function, but it failed completely on the elevator optimization problem, where the PSO variant performed best. "

16.5 Criteria for Choosing an Optimization Algorithm

Direct search algorithms are very popular, for example the Nelder–Mead simplex algorithm or evolutionary algorithms. The popularity of these algorithms is not founded on their overall optimality, but might be related to the following reasons:

1. Direct search algorithms are easy to explain, understand, and implement.
2. For many real-world optimization problems, it is vital to find an improvement, but not the global optimum. Direct search algorithms produce significant improvements during the first stage of their search.
3. Function evaluations are extremely costly in many real-world applications. Hence, the usage of finite-gradient approximation schemes that require at least d function evaluations in every step is prohibitive (d denotes the problem dimension).
4. The analysis of the `sring` optimization reveals that evolution strategies are flexible optimization tools. Tuning can improve their performance significantly.

Direct search algorithms are suitable tools during the first stage of an optimization process. The experimental methodology presented in this chapter provides statistical tools to detect important factors. It provides means for a deepened understanding of the problem, the algorithm, and their interaction as well.

In the experimental setup from this chapter evolution strategies have been demonstrated to be more flexible than PSO, even if the PSO algorithm was combined with modern statistical techniques to cope with the noise. These results extend the analysis from [154] that was based on the noisy sphere function. More generally, we can conclude that simply plugging-in techniques with default parameter settings is not useful: tuning improves performance. A toolbox for the experimental analysis of search heuristics, the so-called *sequential parameter optimization toolbox* [190], is freely available: http://www.springer.com/3-540-32026-1.

On Adaptive Cooperation of AGVs and Elevators in Buildings

17.1 Introduction

Recently, many high-rise buildings taller than 100 m have been constructed, and the concept of "hyper-building", buildings of 1000 m height, has also been proposed and studied [215]. Currently, material handling in such high-rise buildings is carried out mainly by humans using elevators. However, the cost of such vertical transportation for material handling is quite high for the following reason:

- Elevators occupy large floor areas in buildings, to cope with the traffic.
- The size, number of cars, velocity, acceleration, and other factors of the elevators, related to the quality of service, are designed under the constraint of having to carry humans.
- The handling of goods gives rise to labor costs for the humans who take care of them.

We are not aware of much research into measuring the intrabuilding traffic caused by handling small material, such as mail and packets. However, the results of our observation in a building show that more than 50% of the elevator load was caused by such usage [216].

Hence, by automating intrabuilding material handling, it is expected that the quality of transportation service can be improved and that the cost for it is reduced. Further, such systems will make the following advanced services possible:

- High-quality delivery services to support sophisticated businesses.
- Recycling of materials for the protection of the environment.
- Secure handling of materials.
- Delivery services in emergency, such as earthquakes.

Considering this background, the authors have started a research program of studying novel transportation systems for high-rise buildings [216]. This research program consists of studies on the following three subjects:

1. Survey of intrabuilding traffic in existing buildings,
2. Study on modeling and simulation methodologies to treat complex traffic in buildings, and
3. Study on control methods to achieve flexible and robust services.

In this program, the authors have proposed an automatic material handling system consisting of automated guided vehicles and elevators. This chapter describes the progress in our study of the third subject, *i.e.* study on the control of the proposed material handling system. Control strategies based on the concept of autonomous decentralized and multiagent systems are investigated. An intrabuilding traffic simulator for the proposed system has been developed, and control strategies using the contract net protocol (CNP) [217] are applied to it so as to achieve cooperation among agents such as the AGVs and the elevators. An estimator to predict the processing time of carrying an AGV by the elevators has been introduced using neural networks and a learning technique for them, to improve the performance of the CNP. Further, the optimization of the bidding scheme for the CNP using a genetic algorithm has been studied. Numerical simulations show that these methods improve the performance of the system significantly, as compared with a naive heuristic method.

17.2 Material Handling System for High-rise Buildings

The proposed intrabuilding material handling system is schematically illustrated in Fig. 17.1. This system consists of 1) delivery terminals; 2) small AGVs that carry packets; 3) lanes to guide AGVs in horizontal movement; 4) small elevators designed for carrying of AGVs in vertical direction; 5) halls for AGVs to wait for elevator cars.

Once a packet requiring delivery service arrives at a starting terminal, an AGV comes to the terminal and picks it up. Then it rides an elevator to move between floors, if necessary, and carries the packet to the destination terminal.

The proposed system consists of many subsystems such as terminals, AGVs and elevators. A requirement is to provide a flexible and robust delivery service that can cope with changes of system configuration, malfunction of subsystems, change of traffic characteristics, *etc.* Considering these requirements, the authors chose to construct a control system based on the *autonomous decentralized* and *multiagent* concept. That is, each AGV and elevator (in other word, "*agent*") recognizes the situation and decides on appropriate actions by themselves, without central control. Thus, the system can achieve flexible and robust control, coping with changes in its configuration and other factors. In such multiagent systems, the key factor is cooperation among agents. In this study, the authors have adopted the *contract net protocol (CNP)* proposed by Smith [217] to achieve cooperation.

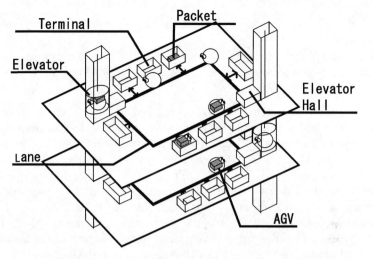

Fig. 17.1. Intrabuilding material handling system

17.3 Contract Net Protocol

To achieve the delivery service with the proposed system, the system must accomplish the following two sorts of tasks:

Delivery service (DS) task: The task of carrying a packet by a AGV from a starting terminal to the destination terminal.

Carrying AGV (CA) task: The task of carrying an AGV by an elevator from the starting floor to the destination floor.

Once a demand for delivering a packet occurs, the terminal selects an appropriate AGV and assigns the *DS* task. The assigned AGV must move to the starting terminal, pick the packet up, and then move to the destination terminal. One *DS* task gives rise usually to two *CA* subtasks of 1) carrying the vacant AGV to the floor of the starting terminal and of 2) carrying the loaded AGV to the floor of the destination terminal. To achieve a *CA* task, the AGV assigns it to an appropriate elevator, and the assigned elevator must send the AGV from the starting floor to the destination of the *CA* task.

For assignment of *DS* and *CA* tasks, the authors adopt the *contract net protocol (CNP)* [217]. The contract net is a model of task assignment through negotiation among agents. In the contract net, an agent who proposes a negotiation is called the *"manager"*, and the candidate agents who may take the task are called the *"contractors"*. The algorithm of the CNP is as follows:

1. **Task Announcement:** As a manager, an agent who has a task to be assigned, announces it to the contractors.
2. **Bidding:** Each contractor selects the executable tasks from the announced ones, and evaluates them. Then, it chooses the task of the highest

priority and sends a bidding message for the task to the corresponding manager.

3. **Awarding:** The manager receives the bidding messages from the contractors, and evaluates them. Then it assigns the task to the contractor of the highest priority.

In the proposed system, the CNP is applied in a two-stage manner. That is, the CNP is utilized for both assignment of the *DS* tasks and the *CA* tasks. Once a demand of delivery occurs, the terminal announces the *DS* task as a manager to the AGVs as contractors. Through negotiation using the CNP, the *DS* task is assigned to a certain AGV. Then the AGV assigns the *CA* subtasks derived from the *DS* task to some elevators, by carrying out negotiation as managers, with the elevators as contractors, see Fig. 17.2.

To achieve efficient control with the multiagent system using the CNP, the most important factor is the function used for evaluation of the announced tasks and bidding. We discuss this in Sections 17.5 and 17.6.

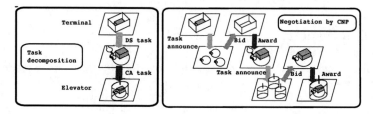

Fig. 17.2. Cooperation among agents in the material handling system using the CNP

17.4 Intrabuilding Traffic Simulator

17.4.1 Outline of the Simulator

To evaluate the control strategies for the proposed system, an intrabuilding traffic simulator is developed. The parameters of the simulated system used in the numerical experiments are shown in Table 17.1a. For the traffic conditions, we used three typical situations shown in Table 17.1b.

17.4.2 Performance Index of Control

The performance of the system should be evaluated considering 1) *Quickness*: The requirement of the service must be processed as quickly as possible. 2) *Equality*: The service must be as equal as possible for each requirement. 3) *Economy*: The resources must be utilized efficiently.

Table 17.1. Outline of traffic simulator

(a) Parameters of the simulated system

Item	Quantity
Simulation time step	1 s
Traffic	Represented by an OD table and average occurrence rate (λ packets/s) of the delivery service in the whole building.
Number of floors	25
Number of terminals	10 terminals/floor
Number of AGVs	100
Speed of AGVs	1/3 s/unit horizontal distance (distance between terminals)
Time of AGV staying at a terminal	4 s
Payload of AGV	1packet/car
Number of elevators	4
Payload of elevator	3 AGVs/car
Time needed for elevator motion	
Stop/Change Directions	1 s
Movement of 1 floor	2 s
Boarding/Leaving of AGV	1 s/ AGV

(b) Traffic conditions

Type	Description	λ
UNIFORM	Origin and destination floors are uniformly distributed	0.16
UP-PEAK	Only the first floor is origin, destination floors are distributed uniformly among the other floors	0.08
DOWN-PEAK	Only the first floor is destination, origin floors are distributed uniformly among the other floors	0.08

In this chapter, since we discuss control with fixed resources such as number of AGVs and elevators, the following performance index to be minimized is used so as to take both quickness and equality into consideration:

$$P = \frac{1}{N} \sum_{i=1}^{N} (T_i)^2$$

where N is the total number of DS tasks, and T_i is the processing time of the ith DS task. Equality of the service is taken into consideration by penalizing long processing time with a quadratic function. Since the simulation is a stochastic process, the value of P fluctuates. A preliminary experiment shows that the performance index P is stabilized enough to allow optimization of the control policy with simulations of 10^5 steps.

17.5 Cooperation based on Estimated Processing Time

17.5.1 Control Using Minimal Processing Time for Bidding

In this section, we discuss a heuristic method of assigning an agent of the minimal estimated processing time through negotiation. In the following, we focus on the negotiation among the AGVs and elevators for assignment of CA tasks with the CNP. As for assignment of DS tasks, we adopt a simple heuristics that each AGV bids using the difference between the floor of the origin terminal and that where the AGV is.

Suppose that each contractor (elevator) can compute the processing time $c(t)$ needed for the announced task t. Then the following heuristic method is expected to work well. That is, the contractor finds a task of the minimal processing time $t_i = \arg\min_i c(t_i)$ among the announced tasks, and sends $c(t)$ as a bidding message b to the manager i of the task t_i. Then the manager finds the contractor of the minimal b among the contractors that sent messages, and awards it.

However, the problem of the above heuristics is that the contractor cannot know the correct $c(t)$. The processing time of an announced task t^{ann} depends on other tasks including tasks that will occur after announcement of the task t^{ann}. Hence, as a simple approximation, we assume the assigned task can be correctly processed, and to bid the task t^{ann} with the minimal processing time $c^-(t^{\mathrm{ann}})$. It can be obtained assuming no new tasks occur later and calculated as a function of the tasks that have occurred so far:

$$c^-(t^{\mathrm{ann}}) = f(T^{\mathrm{cur}} \cup t^{\mathrm{ann}}) \tag{17.1}$$

where T^{cur} is the set of tasks that occurred before the announcement of the task t^{ann}. Since each elevator, say elevator i, can process the tasks already assigned independently, $f(T^{\mathrm{cur}} \cup t^{\mathrm{ann}})$ can be rewritten as $f(T_i^{\mathrm{cur}} \cup t^{\mathrm{ann}})$, where T_i^{cur} is the set of tasks assigned to elevator i. Hence, each elevator can be treated easily by simulating its behavior using the information T_i^{cur} known by it in an autonomous decentralized manner.

17.5.2 Estimation of Processing Time by a Neural Network

Since, $c(t)$ depends on the tasks occurring later, $c(t)$ deviates from the $c^-(t)$. To obtain a more accurate estimation of $c(t)$, a neural network is used as an estimator, which is called "*the neuro estimator*". For estimation of $c(t^{\mathrm{ann}})$, a 3-layer neural network with 10 hidden units is used. As the activation function of the units in the hidden layer, the sigmoid function $y = 1/(1 + e^{-x})$ is used. As for the output layer, the function $y = -0.1 \log((e^{-10x})/(e^{-10x} + 1))$ is used considering the requirement of non-negative output.

To decide the input signals to the neural networks, simulation runs using $c^-(t)$ for bidding are carried out. The result of the analysis of estimation error shows that the estimation error gets larger when the elevator must turn its directions to reach to the boarding floor of the AGV. Considering this, we use the following input signals $u = (u_1, \ldots, u_5)$:

u_1: Whether the elevator must turn its directions to serve the task. If so, $u_1 = 1$, otherwise, $u_1 = 0$.

u_2: Number of CA tasks possessed by the elevator.

u_3: Minimal route length until finishing the task.

u_4: Maximal route length until finishing the task.

u_5: $c^-(t)$.

These input signals are normalized within the interval of $[0, 1)$.

For the learning of the neural network, we have used the conjugate gradient technique for acceleration of convergence, and to reduce the effort for parameter tuning.

17.5.3 Numerical Example

First, with a simulation run using $c^-(t)$ for bidding, we obtain 10 000 sets of u and $c(t)$. With these data, the neural network is trained. Improvement of estimated processing time with the neural network is shown in Fig. 17.3 for the UNIFORM traffic. As shown in Fig. 17.3a, with $c^-(t)$, large estimation error occurs around 150–250 in $c(t)$. Figure 17.3b shows that such error is reduced with the neuroestimator.

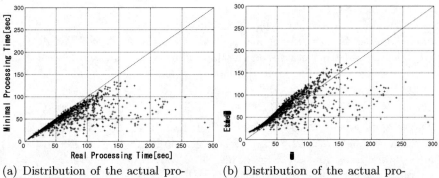

(a) Distribution of the actual processing time $c(t)$ and the minimal processing time $c^-(t)$.

(b) Distribution of the actual processing time $c(t)$ and the estimated processing time $c^e(t)$.

Fig. 17.3. Estimation of processing time

Then, using the estimated processing time $c^e(t)$ with the neural network as bidding scheme, simulation runs are carried out and compared with those using $c^-(t)$. The results are shown in Fig. 17.4. The figures show that the performance of the system is improved remarkably by the neuroestimator both in UNIFORM and UP-PEAK traffic conditions. In the DOWN-PEAK traffic, the improvement is small. The structure of the traffic in this condition is simple, and all the vacant AGVs stay on the first floor. Hence, $c^-(t)$ gives a good estimate of the processing time. This is why the effect of the neuroestimator is limited.

17.6 Optimization of Performance

17.6.1 Bidding Function to be Optimized

In the cooperation method studied in the previous section, the system chooses the contractor with the minimal estimated processing time. Hence, the requirement of equality for each delivery is not considered explicitly, while the quickness of the service in total may be achieved. In this section, in order to achieve equality of the service, as well as quickness, we introduce more factors in evaluation of the announced tasks. For the evaluation function for bidding, the following weighted linear sum of the factors is considered:

$$b = \mathrm{w} \cdot \mathrm{w} \tag{17.2}$$

where $\mathrm{v} = (v_1, v_2, \ldots, v_n)^{\mathrm{T}}$ are the factors and $\mathrm{w} = (w_1, w_2, \ldots, w_n)^{\mathrm{T}}$ is the weight vector w, which is to be optimized. As the input vector w, the following seven signals $(\mathrm{w} = (v_1, \ldots, v_7))$ are used:

v_1: Directions of the elevators and CA task. If they are the same, $v_1 = 1$, and otherwise 0.

v_2: Number of CA tasks assigned to the elevator.

v_3: Time needed for staying at the floor where the AGV will board.

v_4: Time needed for staying at the floor where the AGV will leave.

v_5: Distance between current floor and the main floor.

v_6: Square of the estimated processing time $(c^{\mathrm{e}}(t))^2$.

v_7: The estimated processing time $c^{\mathrm{e}}(t)$.

These input signals are normalized within the interval of $[0, 1)$. Initial values of the weights are taken randomly following a uniform distribution on $(-1, 1)$ for v_1 through v_5, and $(0, 1)$ for v_6 and v_7.

Since the estimated processing time $c^{\mathrm{e}}(t)$ is included in the input signals, and the evaluation function, Equation 17.2 is a linear one, the search space of the evaluation function includes the bidding using $c^{\mathrm{e}}(t)$ discussed in the previous section, and further improvement of the performance is expected by optimization of the evaluation function.

17.6.2 Application of Genetic Algorithm

Since the performance index obtained by simulation has fluctuations and no gradient information is available, the authors adopt a real-coded genetic algorithm, a genetic algorithm using a floating-point genetic representation, for optimization. The MGG model proposed by Satoh *et al.* [218] and the unimodal normal distribution crossover (UNDX) proposed by Ono *et al.* [78] are used as a generation alternation model and a crossover operator, respectively. Since the UNDX has excellent search ability by itself, no mutation is used. With a simulation run of 10^5 steps using w for the bidding scheme, the performance index P for w is measured and used for the fitness function to be minimized. Since fitness evaluation requires time-consuming simulation of the material handling system, the population size is set at 40, the number of children generated in MGG is set at 2, and the maximal number of fitness evaluations is set at 2000, based on a preliminary study.

As stated in Section 17.4, the traffic simulation for fitness evaluation uses random numbers. If a fixed series of random numbers is used, undesirable solutions adapted to the series may be obtained. On the other hand, if each simulation uses a different series, the optimization gets difficult because of the large variance of the fitness evaluations. Hence, we select five series of random numbers in advance, and one of them is chosen randomly and applied to a single simulation run for fitness evaluation.

17.6.3 Numerical Example

Optimization by the GA is applied for the traffic conditions of UNIFORM, UP-PEAK and DOWN-PEAK. For each traffic condition, 5 trials of GA are carried out. With the obtained weight vector, 100 simulation runs are made to measure the distribution of the performance, and compared with those obtained by the neuro estimator discussed in the previous section. The performances of the obtained solutions are shown in Fig. 17.5.

These figures show that with optimization using the GA, further improvement of the performance is achieved compared with the heuristic method using the neuroestimator. In UNIFORM and DOWN-PEAK traffic conditions, the improvement of the performance is remarkable. In UP-PEAK traffic, however, the improvement stays small. Examining the effects of obtained weights with sensitivity analysis, we find that the inputs v_2 and v_3 play an important role, as well as v_7 in the UNIFORM traffic. That is, the system considers whether the elevator stops at the origin or the destination floor to serve some of the already assigned tasks. Further, v_6 also contributes to the decision making. For UP-PEAK and DOWN-PEAK traffic conditions, the contribution of v_2 and v_3 gets smaller, and v_5, the distance from the first floor to the elevator, contributes more.

17.7 Conclusion

In this chapter, autonomous decentralized control strategies for intrabuilding material handling system consisting of agents such as terminals, AGVs and elevators, are discussed. We adopt the contract net protocol (CNP) as

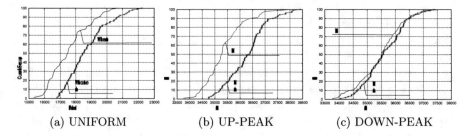

(a) UNIFORM   (b) UP-PEAK   (c) DOWN-PEAK

Fig. 17.4. Effect of the neuroestimator. The abscissa represents the performance index P, and the ordinate represents the cumulative frequency, *i.e.* the number of trials showing performance less than or equal to the corresponding P in 100 trials.

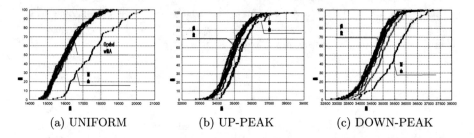

(a) UNIFORM   (b) UP-PEAK   (c) DOWN-PEAK

Fig. 17.5. Results of optimization with the GA

a fundamental mechanism to achieve cooperation among agents. The control system works well with a simple heuristics of using minimal processing time of the elevator as the bidding scheme in the CNP. It is improved by introducing an estimator with a neural network. Further improvement is achieved by optimizing the bidding scheme by a genetic algorithms. Improving the cooperation among terminals and AGVs, adaptation to change of traffic, and functional specialization are the subjects of future study.

Optimal Control of Multicar Elevator Systems by Genetic Algorithms

18.1 Introduction

Elevator systems are the most important transportation method for high-rise buildings. However, in the conventional elevator, only one elevator car, suspended by ropes, occupies the whole elevator shaft. Because of this, the taller the building is, the lower the performance of the elevator becomes. With the background of progress in linear-motor technology and increasing needs for high-performance transportation systems for large-scale buildings, rope-less multicar elevators (MCE) that have several cars driven by linear motors in a single elevator shaft, are attracting attention as a novel transportation system [75, 219, 220]. However, considering the control of MCE, we can only apply the knowledge of existing elevator systems to MCE to a limited extent, and there is a need for new controller design methods. Simulation-based optimization of the control strategy is one of the candidates for the design methods. Sudo *et al.* [220] have demonstrated that a genetic algorithm can find good control strategies through simulation-based optimization for MCE systems. However, it takes a very long computation time because

1. Evaluation of a control strategy is obtained through complex discrete event simulation of the MCE system, and
2. Multiple simulation runs are needed to reduce random fluctuation in the evaluation results, due to the fact that the simulation uses random numbers.

This chapter discusses acceleration of the simulation-based optimization of an MCE controller. We have adopted two devices for acceleration. One is a genetic algorithm explicitly considering fitness functions involving noise. The other is utilization of a PC cluster system for parallel evaluation of fitness values. With these devices, optimization of the controller and evaluation of the obtained results are carried out within a practical computation time, and it enables detailed investigation of a control scheme for MCE.

This chapter is organized as follows. In Section 18.2, we provide a brief overview of the considered MCE system and simulation-based optimization of the controller for it . Section 18.3 explains the *memory-based fitness evaluation genetic algorithm* (MFEGA) proposed by Sano *et al.* [106, 107, 129] for fast optimization of a noisy fitness function. Section 18.4 is a comparative study of GAs for optimization of the controller for MCE. In Section 18.5, some design parameters of the controller are examined using the proposed method. Section 18.6 is the conclusion of this chapter.

18.2 Multicar Elevator Systems and Controller Optimization

18.2.1 Multicar Elevator Systems

The MCE system, considered in this chapter consists of the following elements, see Fig. 18.1.

Floors: We assume that the floor having the entrance and the exit of the building is located at the lowest level of the building, and it has the largest traffic. We call it *the terminal floor*. The other floors are assumed to be identical in their traffic demand, and we call them *the general floors*.

Elevator shafts: They are the space where elevator cars move. We assume the considered building has 6 shafts based on the experience of existing buildings and a preliminary study on MCE.

Elevator cars: We assume that each elevator shaft has two cars. As for the configuration of the MCE system, we assume that the elevator car can move only vertically in the corresponding elevator shaft taking technical feasibility, safety and comfort of users into consideration.

However, in this type of MCE, efficient control such as assignment of the traffic demand to a certain car is difficult because one car can't pass the other car in the same shaft. This is the motivation of our study of searching for a good control strategy, using an optimization technique.

Further, we have assumed the following as the configuration of the controller for MCE:

Zone operation: To avoid collision of cars and ease of operation, we introduced *zone operation*. That is, the floors are divided into upper and lower zones. The upper car in each shaft serves traffic demands whose origin or destination is in the upper zone. The lower car serves traffic only in the lower zone.

Garage floor: To allow the upper car to serve the terminal floor, a *garage floor* is introduced below the terminal floor, where the lower car can wait.

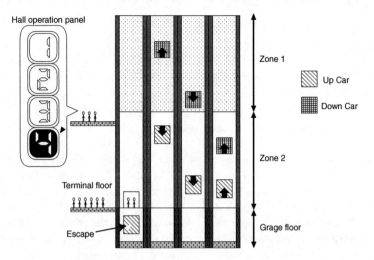

Fig. 18.1. Multicar elevator (MCE) system

Registration of destination floor: The users are asked to register their destination floors at the halls. It is needed to guide the users to appropriate
cars, according to the *zone operation*.

Collision avoidance: To avoid collision of cars and deadlock, the cars in a shaft
are not allowed to move towards each other.

18.2.2 Controllers for MCE

Control of MCE is carried out by assigning a destination call to a certain
car. This is decided according to the definition of the service zone, and the
assignment policy. If a new hall call is registered, each shaft is requested to
find a car that can serve the call. Then, all such cars are evaluated by the
weighted sum of 11 attributes expressing the state of the car, such as the
number of calls already assigned to the car, and the expected arrival time.
The car having the minimum weighted sum of attributes $\{v_1, \cdots, v_{11}\}$ with
weights $\{w_1, \cdots, w_{11}\}$ is assigned to the call:

$$\min \sum_{i=1}^{11} w_i v_i \tag{18.1}$$

18.2.3 Discrete Event Simulation of MCE

For simulation of the MCE, we have used a simulator developed by the authors. The simulator is based on the discrete event model called the *extended
state machine* (ESM), introduced in Chapter 7, which models the system by

finite state machines that have timers and send and receive messages mutually for synchronization [75, 216].

In the ESM model, each of the elevator cars and their doors is represented by an ESM. Further, a passenger is also represented by an ESM, which is dynamically generated and deleted.

18.2.4 Simulation-based Optimization

In this study, we consider a building with the specification shown in Table 18.1. The weight vector of the controller for the MCE is optimized through discrete event simulation, using genetic algorithms (GA). The configuration of the GA is shown in Table 18.2.

In our formulation shown in Equation (18.1), the weight vector has a redundancy, since the control behavior is the same, even if it is multiplied by a positive constant. Hence, we have fixed the value of w_1 at 0.5, because it is expected to be positive, following from the definition of attribute v_1, which is the expected arrival time. Initial values of other weight elements are randomized, according to a uniform distribution over $(-1, 1)$.

The performance of a weight vector is evaluated by the average square waiting time of passengers. To reduce the effect of the transitional stage of the traffic, simulation results in an initial interval are excluded from evaluation. Elevator traffic is represented by an OD distribution. In our study, we use the following OD distribution: (Terminal Floor $\leftrightarrow$ General Floor : General Floor $\leftrightarrow$ General Floor) = (10 : 1).

Table 18.1. Specification of the elevator system

Item	Value
No. of floors	30
Floor height	4.34 m
dv^2/d^2t of car	2 m/s^3
Maximum car acceleration	1.1 m/s^2
Maximum load	20 persons/car
Time needed for	
Door opening	1.8 s
Door closing	2.4 s
Entering/leaving	1.2 s/person
No. of elevator shafts	6
No. of cars/shaft	2
Traffic	2700 persons/h

Table 18.2. Configuration of genetic algorithm

Item	Value
Individual	Weight vector
Fitness	Mean sq. waiting time
Generation alternation	MGG model [78]
Crossover	UNDX [78]
Pop. size	35
No. of children/gen.	6

18.2.5 Problems in Optimization

In the optimization of the controller for MCE, we found the following difficulties, due to the simulation-based optimization:

Computation time: In the application of GA to the optimization of MCE controller by Sudo *et al.*, a single GA run requires more than 1 week because a single simulation run takes more than 30 s.

Fluctuation of evaluation: Since the discrete event simulation uses random numbers, the fitness value fluctuates. This causes the optimization to be difficult and inaccurate. In the previous study, the effect of randomness is reduced by taking the average of five simulation runs; but that takes a longer computation time.

Control scheme: Because of computation time, the examination of various control schemes, such as different zone settings, and extension of attributes used for control, becomes difficult.

18.2.6 Acceleration of Computation

In the current work, acceleration of computation is achieved by adopting the following devices:

Usage of GA for Noisy Fitness Function

For optimization of the fitness function involving noise, we adopt a more sophisticated variation of GA, that takes it explicitly into consideration, instead of the simple approach of using average performance of several simulation runs.

Usage of PC Cluster

In GA, several solution candidates are evaluated in parallel, and in simulation-based optimization of the MCE controller, one simulation run takes a long time. Hence, we adopt a PC cluster for parallel evaluation. We use a PC cluster with 6 PCs, each with dual processors; and we adopt a master–slave architecture.

18.2.7 Re-examination of Configuration of Simulation

The longer time we allow for simulation, the more accurate the observed performance becomes. However, simulation time that yields sufficiently stable results depends largely on the configuration of the MCE. We examined the time spent on each simulation run through a preliminary study.

18.3 A Genetic Algorithm for Noisy Fitness Function

In a previous study, the influence of random fluctuation in the fitness value is reduced, by averaging five simulation runs using a prescribed set of random sequences. In the following, we call it Sample-5 GA.

However, this method faces the difficulty of a heavy computational load. In this work, we adopt the *memory-based fitness evaluation GA (MFEGA)* proposed by Sano *et al.* [106, 107, 129] so as to reduce the computation. In MFEGA, an optimization problem with a noisy fitness function is formulated as follows:

$$\min_{\boldsymbol{x}} \langle F(\boldsymbol{x}) \rangle \tag{18.2}$$

$$F(\boldsymbol{x}) = f(\boldsymbol{x}) + \delta \tag{18.3}$$

where $f(\boldsymbol{x})$ is the fitness value, δ is additive noise, and $\langle \rangle$ is the expectation.

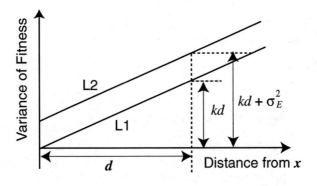

Fig. 18.2. A stochastic model of noisy fitness

In the MFEGA, all the individuals generated so far are stored in memory, with the observed fitness values as a history, and the fitness value of a novel search point is estimated by the fitness values in the history, as well as by the sampled fitness value at that point. For this purpose, MFEGA employs a simple stochastic model of fitness value shown in Fig. 18.2. Let $\boldsymbol{x}$ and $f(\boldsymbol{x})$ be the point of interest, and the fitness value at that point, respectively. Let $f(\boldsymbol{h})$

be the fitness value at h whose distance from x is d. Then MFEGA assumes a model

$$f(h) \sim N(f(x), kd) \tag{18.4}$$
$$d = ||h - x|| \tag{18.5}$$

where k is a positive constant. Assuming that the additive noise δ follows a normal distribution with variance σ_E^2, we obtain

$$F(h) \sim N(f(x), kd + \sigma_E^2) \tag{18.6}$$

The fitness value $f(x)$ is estimated with this model. Let $h_l = 1, \cdots, H$ be the individuals in the history H, and let $F(h_l)$ and d_l be the sampled fitness value at h_l and distance from x in the history, respectively. The probability of obtaining $F(h_l), l = 1, \cdots, H$ is given by

$$\prod_{l=1}^{H} p(F(h_l), d_l) \tag{18.7}$$

where $p(F(h_l), d_l)$ is the p.d.f. (probability density function) of $F(h_l)$, and it is given by

$$p(F(h_l), d_l) = \\ \frac{1}{\sqrt{2\pi(k'd_l + 1)\sigma_E^2}} \exp\left(-\frac{1}{2} \frac{(F(h_l) - f(x))}{(k'd_l + 1)\sigma_E^2} \right) \tag{18.8}$$

With the maximum likelihood estimation taking Equation (18.8) as the likelihood of $f(x)$, we obtain an estimation of $f(x)$ as a weighted sum of the sampled fitness values as follows:

$$\tilde{f}(x) = \frac{\sum_{l \in H} \frac{F(h_l)}{k'd+1}}{\sum_{l \in H} \frac{1}{k'd_l+1}} \tag{18.9}$$

where $k' = k/\sigma_E^2$ is a parameter of MFEGA.

In the MFEGA, the selection operation of conventional GA is carried out by using the estimated fitness value given by Equation (18.9).

18.4 Comparison of GAs for Noisy Fitness

18.4.1 Setup of Experiments

In this section, we evaluate the MFEGA through a comparative study. We compared

1. Standard GA: GA without particular consideration of noise.
2. Sample-5 GA: GA using average of the fitness values in 5 simulation runs with 5 fixed random sequences.
3. MFEGA.

We have applied these methods to the optimization of a controller for a 6-shaft 2-car MCE with the zone boundary (the lowest floor of the upper zone) at 16. Each GA run is carried out up to 400 generations. For each method, we carried out 3 GA runs, and for each solutions obtained by GA, its performance is evaluated by sampling fitness values in 100 simulation runs using different random sequences.

18.4.2 Results of Experiment

The distributions of fitness values are shown in Figs.18.3a – c, and the statistics of performance of the obtained solutions are summarized in Table 18.3. As for the quality of obtained solutions, Sample-5 GA and MFEGA show similar performance and it is better than the performance of standard GA. This shows that these GAs find good solutions even with large random fluctuations of fitness. As for computation time, however, MFEGA takes a similar time to the standard GA, and it is much faster than the Sample-5 GA that requires 5 times longer simulation runs. Considering these experimental results, we adopted the MFEGA in the following experiments.

Table 18.3. Statistics of performance of GAs

Type of GA	Performance	
	Mean	Std. dev.
Standard GA		
Trial 1	2747	231
Trial 2	2782	297
Trial 3	2783	244
Sample-5 GA		
Trial 1	2581	268
Trial 2	2628	238
Trial 3	2639	258
MFEGA		
Trial 1	2624	265
Trial 2	2639	254
Trial 3	2783	373

For each GA, trials are sorted by mean performance.

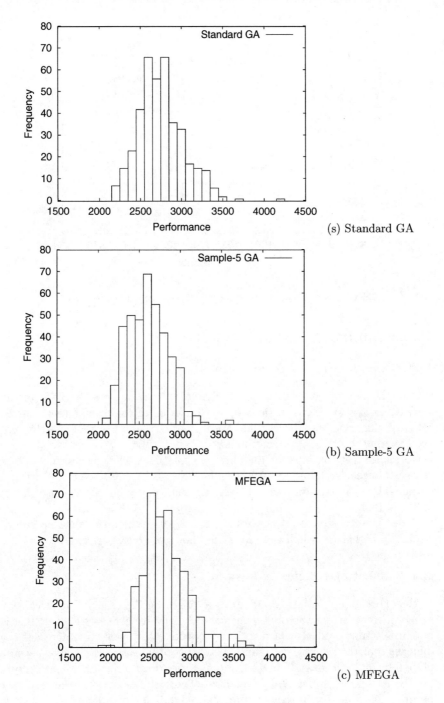

Fig. 18.3. Comparison of genetic algorithms

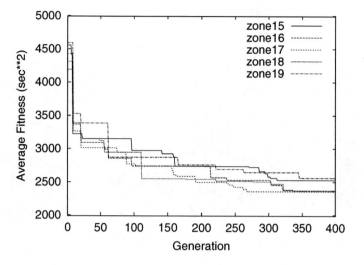

Fig. 18.4. Analysis of zone partition (evolution process)

18.5 Examination of Control Strategy

18.5.1 Examination of Zone Boundary

One of the important parameters in the control strategy is the floor of the zone boundary. We have carried out optimization of the weight vector using MFEGA, for zone boundaries (the lowest floor of the upper zone) 15 through 19. The time course of optimization is shown in Fig. 18.4. Figures 18.5a – e are performance distributions for the obtained solutions. They are obtained by 2 GA runs for each zoning, and 150 simulation runs to evaluate each controller obtained by GA. The statistics of the performance of solutions are also shown in Table 18.4.

These results show that the zoning of *zone-17*, where the zone boundary is slightly higher than halfway (*zone-16*) shows the best performance.

18.5.2 Effect of Weight Extension

In the experiments so far, the weight vectors of all the cars are identical. By using different weight vectors for the cars, the performance of the controller is potentially improved, if it is sufficiently optimized. However, it increases the dimension of the search space, and therefore optimization itself may become more difficult. In this section, we examined the use of different weight vectors for the groups of upper and lower cars. Thus, the decision variable vector becomes 22-dimensional (considering the redundancy mentioned earlier, it is 20-dimensional). For this problem, the crossover operator is applied to the weight vectors for upper and lower cars separately.

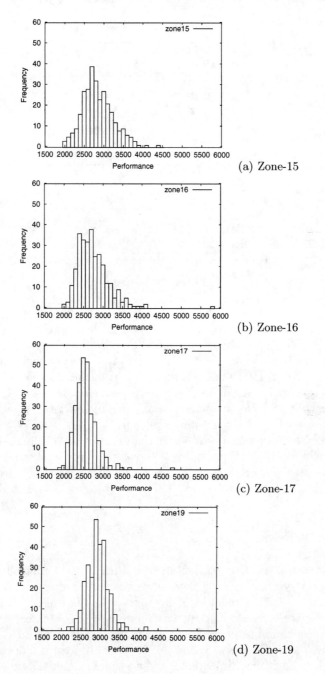

(a) Zone-15

(b) Zone-16

(c) Zone-17

(d) Zone-19

Fig. 18.5. Analysis of zone partition (distribution of performance)

Table 18.4. Statistics of performance of zoning

Zoning	Performance	
	Mean	Std. dev.
Zone-15		
Trial 1	2786	345
Trial 2	2797	335
Zone-16		
Trial 1	2706	471
Trial 2	2794	353
Zone-17		
Trial 1	2525	242
Trial 2	2588	348
Zone-19		
Trial 1	2801	360
Trial 2	2947	281

For each zoning, trials are sorted by mean performance.

Time courses of 3 GA runs in the 22-dimensional problem are shown in Fig. 18.6 as well as one trial in the original 11-dimensional problem. The distribution of performances obtained by the three GA runs are shown in Fig. 18.7. For each controller, 150 simulation runs are carried out. As shown in Fig. 18.6, the 22-dimensional problem shows poorer performance in the beginning of optimization, but in 2 cases among 3 trials, they converge to solutions better than a trial for the 11 dimensional problem. However, Fig. 18.7 shows that the obtained performance is still worse than that of the 11-dimensional problem shown in Fig 18.3c. Considering the evolution process shown in Fig. 18.6, better solutions may be obtained in the 22-dimensional problem by extending the number of generations for GA. This is a subject of future study.

18.6 Conclusion

This chapter examined simulation-based optimization of controllers for multicar elevator systems, using genetic algorithms. To reduce the computation time and the influence of random fluctuation in the fitness value obtained through discrete event simulation, we employed *MFEGA*, a genetic algorithm for fast optimization of noisy fitness function; and we used a PC cluster for parallel evaluation of fitness values. With these devices, the optimization is completed in a practical time frame, and we could examine optimal zone partitioning, and the effect of extending the decision variables for the controller of the MCE.

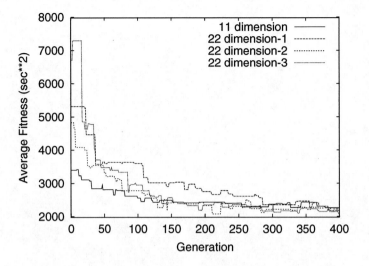

Fig. 18.6. Evolution process in the 22-dimensional formulation

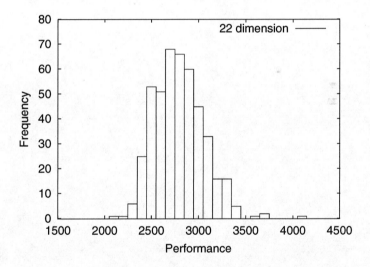

Fig. 18.7. Performance in the 22-dimensional formulation

Analysis and Optimization for Automated Vehicle Routing

19.1 Introduction

This chapter deals with an automated material handling system (MHS) referred to as a permutation circulation-type vehicle-routing system (denoted PCVRS). It consists of a fleet of (robotic) vehicles, multiple stations including an input/output one and a guide path of a single loop (see Fig. 19.1). Each vehicle unidirectionally and repeatedly circulates on the loop to serve jobs at stations located along the loop. In the PCVRS no passing is allowed between vehicles on the loop (from which the system is named a permutation circulation-type). We can find such PCVRSs in many real MHSs in automated storage and retrieval systems (AS/RSs) and flexible manufacturing systems (FMSs) among others.

The main reason for the practical use of the PCVRS is the simplicity in design and control [221]. An important issue of the PCVRS is the interference (or blocking) between vehicles that occurs when a vehicle is forced to stop by the preceding one for collision-avoidance. Such interference induces wasteful waiting time and wasteful energy consumptions by deceleration and acceleration, and may deteriorate system performance. Thus, many previous papers on the PCVRS avoid the interference by taking a sufficient distance between vehicles. However, the optimal vehicle routing that we consider here does not always mean the same distance between vehicles as assumed in many previous papers and a PCVRS that guarantees interference-free operation may not be realistic when the number of vehicles is increased under a fixed loop length in order to increase the throughput. For example, in big automated warehouses (by which our study was motivated) a lot of items simultaneously supplied by big trucks from the outside have to be stored in a short time (*e.g.* a couple of hours) and a lot of items stored have to be shipped out in a very short time. In these situations high throughput is desired even if interferences are unavoidable.

The objective of this chapter is to discuss vehicle-routing rules for minimizing the interference and maximizing the throughput. A vehicle-routing rule decides the stations which each vehicle in each lap serves.

This chapter is organized as follows. Section 19.2 describes basic assumptions for "parallel and bottleneck-free" PCVRS. It discusses the dynamics of the PCVRS with infinite acceleration and deceleration of vehicles and introduces the "steady state" and the "one lap behind" (denoted OLB) interference, and introduces the "throughput" and the "mean interference time" as performance measures. Section 19.3 explains two basic vehicle-routing rules (random rule and order rule) introduced in the literature and applies the results in Section 3 to them. Section 19.4 introduces two vehicle-routing rules ("E-order" rule and "D-order" rule) and shows that the E-order rule is optimal in the steady state, but the D-order rule is better than the E-order rule under the OLB interference. Section 19.5 reports results obtained by simulation and shows that they confirm the theoretical results. Section 19.6 contains concluding remarks.

19.2 Basic Assumptions and Basic Analysis

This section provides some basic assumptions on the PCVRS and basic analysis on the dynamics of the PCVRS.

19.2.1 Parallel and Bottleneck-Free PCVRS

As shown in Fig. 19.1, the PCVRS consists of a loop of length L, a set of n_V identical vehicles $V = \{V_1, V_2, \ldots, V_{n_V}\}$ (the vehicles are indexed in the order of their circulation), a single input/output station (denoted I/O or S_0) where items are loaded or unloaded by vehicles, and a set of n_S processing stations $SP = \{S_1, S_2, \ldots, S_{n_S}\}$ (the stations are indexed in the clockwise direction). It is implied by a saving job that a vehicle receives a (unit) item at station I/O, sends it to a certain processing station, unloads it there and then returns to station I/O empty. It is implied by a retrieval job that an empty vehicle goes to a certain processing station, loads an item there, gets back to station I/O and outputs it there. This exclusiveness of these two jobs is more effective than a job that includes both saving and retrieving operations for avoiding interferences with a big fleet of vehicles, and hence is found in many real systems such as AS/RS. In the following we assume only the saving jobs, but the results obtained hold also for retrieving jobs.

There is a set of n_J (saving) jobs, $J = \{J_1, J_2, \ldots, J_{n_J}\}$ to be served in a given planning horizon, where the jobs are indexed according to the input order at station I/O. In the following we assume that every vehicle serves a job in each lap, *i.e.* no empty circulation is allowed. Let $V_{k(i)}$ be the vehicle serving job J_i, then

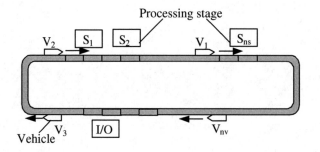

Fig. 19.1. Permutation circulation-type vehicle-routing system, PCVRS

$$k(i) = i - \lfloor (i-1)/n_V \rfloor n_V, i = 1, 2, \ldots, n_J \qquad (19.1)$$

where $\lfloor x \rfloor$ stands for the largest integer not greater than x [10]. We assume that all processing stations (excluding I/O) can equally serve any job to keep the load balance. This means a parallel system that could, in general, give the highest system efficiency and the highest system reliability, and means that processing time p_P on any station of SP and p_0 on loading or unloading station are, repectively, constant, *i.e.* let $p_m(i)$ be the processing time of job J_i on station S_m, then

$$p_m(i) = p_P > 0 \; m = 1, 2, \ldots, n_S, \; i = 1, 2, \ldots, n_J, \text{ if } S_m \text{ really serves } J_i$$
$$p_m(i) = p_0 > 0 \qquad\qquad\qquad m = 0, i = 1, 2, \ldots, n_J$$
$$p_m(i) = 0 \qquad\qquad\qquad\qquad \text{else} \qquad\qquad (19.2)$$

We assume that station I/O is never a bottleneck, meaning that

$$p_0 \le p_P/n_S \qquad (19.3)$$

The constant processing times are realistic in many AS/RSs, when each vehicle carries a unit load. We also assume in the following that the number of jobs is a multiple of the one of the vehicles, unless stated otherwise.

19.2.2 Interferences and Steady State

This section assumes that the acceleration and the deceleration of the vehicle are infinite and every distance between any two points is measured by time units a vehicle takes to run without stop. It also assumes that every vehicle runs with the same and constant speed and hence the corresponding metric distance is obtained by multiplying by the speed of the vehicle. We employ the so called zone control policy that keeps the distance between two adjacent vehicles at least d_B for collision avoidance and assume that the distance between every two adjacent vehicles is d_B when they leave at the I/O station

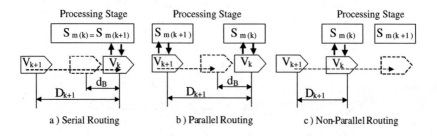

Fig. 19.2. Types of routings

in the first lap (referred to as the minimum start interval). Let D_{k+1} be a distance from vehicle V_{k+1} to its preceding vehicle $V_k (k = 1, 2, \ldots, n_V)$, where D_{n_V+1} stands for one from V_1 to V_{n_V} (in the clockwise direction). Obviously,

$$L = \sum_{k=1}^{n_V} D_{k+1} \tag{19.4}$$

and

$$\Delta_{k+1} \equiv D_{k+1} - d_B \geq 0 \tag{19.5}$$

Let $S_{m(k)}$ be a station where V_k serves a job in a certain lap, then the routings of two vehicles V_k and V_{k+1} are called serial if $m(k) = m(k+1)$, parallel if $m(k) > m(k+1)$, and nonparallel if $m(k) < m(k+1)$ (see Fig. 19.1 and Fig. 19.2). Note that the routing on station I/O for every vehicle is serial. V_{k+1} is interferred (or blocked) by V_k in nonparallel or serial routing, if V_k stays p time units on a station (e.g. $p = p_P$ on a processing station and $p = p_0$ on the I/O one) and

$$\Delta_{k+1} < p \tag{19.6}$$

then V_{k+1} has to wait behind V_k for a time given by

$$W_{k+1} \equiv p - \Delta_{k+1} = p + d_B - D_{k+1} \tag{19.7}$$

In other words V_{k+1} is not blocked if

$$D_{k+1} \geq D_{k+1}^* \equiv p_P + d_B \text{ on a processing station} \tag{19.8}$$

$$D_{k+1} \geq D_{k+1}^* \equiv p_0 + d_B \text{ on station I/O} \tag{19.9}$$

The interference in the serial routing is referred to as serial and the one in the nonparallel routing to as nonparallel. Note that under the assumption of constant processing times (Equation (19.2)) no interference in parallel routing occurs. An interference with a vehicle may be infectious to the successors. Vehicle $V_{k+l}(l = 2, 3, \ldots, n_V - k)$ waits for

$$W_{k+l} = \max\{0, W_{k+l-1} - \Delta_{k+l}\} = \max\{0, p - \sum_{h=1}^{l} \Delta_{k+h}\} \quad (19.10)$$

where $\Delta_{k+l} = D_{k+l} - d_B$ [10]. Let D'_{k+1} be the distance from V_{k+1} to V_k after their serial or nonparallel services in a certain lap, then

$$D'_{k+1} = \max\{p + d_B, D_{k+1}\} \geq D_{k+1} \quad (19.11)$$

Thus V_{k+1} is not blocked on the same station unless V_k is interferred afterward. Let $D_{k+1} = D^*_{k+1}$ be the minimum interference-free (denoted MIF) distance with which V_{k+1} is not blocked by V_k (by Equations (19.8) and (19.9)), and

$$L^*_V \equiv D^*_1 + D^*_2 + \ldots + D^*_{n_V} \quad (19.12)$$

be the minimum fleet length (denoted MFL) of the n_V vehicles that guarantees interference-free operation (see Equation (19.4)). The fleet of n_V vehicles with the MFL is divided into n_{sg} subgroups, each having n_{sw} vehicles except the last subgroup as shown in Fig. 19.3 [222]. The distance between two adjacent vehicles within a subgroup is $p_0 + d_B$, and the one between two adjacent subgroups is $p_P + d_B$. n_{sg} and n_{sw} depend on the vehicle-routing rule used as discussed in the next sections. This fact may eventually lead to a state referred to as a (deterministic) steady state in which no interference at any station occurs in the remaining laps (a state before steady state is referred to as transient). This means that the vehicles should circulate in a steady state from the first lap, if possible. However, any steady state may be impossible when the one lap behind interference occurs as discussed in the following.

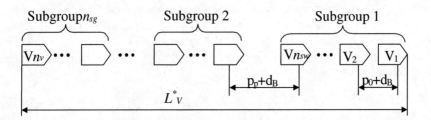

Fig. 19.3. Minimum fleet length MFL in steady state

19.2.3 One Lap Behind Interference

We say that the one lap behind (denoted OLB) occurs, if V_1 is blocked by V_{n_V} (the last vehicle in the latest lap) with the MFL. That is, the OLB occurs on a processing station, if

$$L - (L_V^* - p) < D_{n_V+1}^* = p_P + d_B \tag{19.13}$$

or on the I/O station if

$$L - (L_V^* - p) < D_{n_V+1}^* = p_0 + d_B \tag{19.14}$$

where L satisfies (19.4) and p stands for processing time of V_1 on station $S_{m(1)}$ on which V_1 serves immediately before the OLB, (*i.e.* $p = p_0$ for on station I/O, $p = p_P$ on every processing station). Note that when V_1 is blocked on a station, the MFL is reduced to $L_V^* - p$ by its service on the immediately preceding station. When the vehicles start with the minimum start interval d_B on station I/O, MIF distances (19.8) and (19.9) and the MFL (19.12) are eventually realized by Equation (19.11) unless the OLB interference occurs, but it does not mean that once a vehicle realizes the MIF distance the vehicle keeps it in the remaining laps. For example, even if $D_{k+1} = D_{k+1}^*$ is realized in a certain lap, $D_{k+1} < D_{k+1}^*$ occurs when V_k is blocked by its preceding vehicles in the following laps (*e.g.* by infection, Equation (19.10)). However, once V_{k+1} takes the MIF distance after every preceding vehicle takes its MIF distance, then V_{k+1} is never disturbed in the remaining laps unless the OLB interference occurs. This means that the vehicles reach a steady state, according to (19.12),(19.13) and (19.3), if and only if

$$L - L_V^* \geq p_P + d_B - p_0 \tag{19.15}$$

otherwise, they suffer from interferences including the OLB one in every lap [11].

19.2.4 Throughput and Mean Interference Time

One of the most important performance measures for the PCVRS is the average throughput rate (simply referred to as the throughput) that is defined by the average number of jobs output from station I/O per unit time. Let $F_{\max}$ be the time period to process the n_J jobs (*i.e.* makespan), then the throughput is defined by

$$T_p \equiv n_J / F_{\max} \tag{19.16}$$

Let the n_V vehicles take n_J/n_V laps to process n_J jobs, then,

$$F_{\max} \geq (n_J/n_V)(L + p_P + p_0)$$

by neglecting the interference time. Thus,

$$T_p \leq T_p^V \equiv \frac{n_V}{L + p_P + p_0} \tag{19.17}$$

T_p^V is a vehicle-based upper bound that linearly increases with n_V [11]. If no OLB interference occurs,

$$\lim_{n_J \to \infty} T_p = T_p^V \tag{19.18}$$

Let L_{n_S} be the time units for a vehicle to move from S_1 to S_{n_S} and consider a situation in which the n_S stations are always busy to serve n_S jobs except $(L_{n_S} + d_B)$ time units to simultaneously exchange the next n_S vehicles that wait behind S_1. This situation makes the throughput maximum, then

$$T_p^S \equiv n_S / (p_P + L_{n_S} + d_B) \tag{19.19}$$

is a station-based upper bound that increases with n_S. Thus

$$UB_{T_p} \equiv min\{T_p^V, T_p^S\} \tag{19.20}$$

is an upper bound of the throughput. Another important measure is the total flow time (denoted TFT) that is the total sum of traveling time over n_J jobs or equivalently the mean flow time per job (denoted MFT). TFT and MFT are, respectively, given by

$$TFT = (L + p_P + p_0)n_J + TIT$$
$$MFT = L + p_P + p_0 + MIT \tag{19.21}$$

where TIT and MIT, respectively, stand for the total interference time over n_J jobs and the mean interference time per job. Thus to minimize TIT (MIT) is equivalent to minimizing TFT (MFT). So the MIT is used for the evaluation in the following. Let the fleet of vehicles take the minimum start interval in the first lap, then the distance from V_1 to V_{k+1} is expanded (by interferences) from $(k-1)d_B$ to $(D_2^* + \ldots + D_{k+1}^*)$ when it reaches a steady state (see Fig. 19.3). Then,

$$MIT^* \equiv \frac{\sum_{k=2}^{n_V}(n_V - k + 1)(D_k^* - d_B)}{n_J^*} \tag{19.22}$$

is the MIT with the minimum start interval, where n_J^* stands for the number of jobs with which the steady state is realized. Let n_V^* be the maximum number of vehicles with no OLB interference and T_p^* be the maximum throughput that is realized by n_v^* in a steady state. The following two sections will show that these performance measures strongly depend on the vehicle-routing rules employed.

19.3 Two Basic Vehicle Routings

This section analyzes two basic vehicle-routing rules that keep the load balance to all processing stations for the parallel and bottleneck-free system to be effective (see Section 19.2.1).

19.3.1 Random Rule

A vehicle-routing rule is referred to as *random*, if every vehicle serves a job on a processing station at random with the same probability [221]. Every two vehicles take serial and/or nonparallel routing under the random rule. Then, when a steady state is reached, although it is stochastic, the vehicles are spaced equally apart at intervals according to Equation (19.8),

$$D^R_{k+1} \equiv D^*_{k+1} = p_P + d_B, \ k = 1, 2, \ldots, n_V - 1$$

and the MFL becomes, according to Equation (19.12)

$$L^R_V \equiv L^*_V = (n_V - 1)(p_P + d_B) \tag{19.23}$$

thus, the steady state is realized by Equation (19.14), if

$$L \geq n_V(p_P + d_B) - p_0 \tag{19.24}$$

The maximum number of vehicles satisfying Equation (19.24), the vehicle-based upper bound of the throughput derived from Equation (19.16) are given by

$$n^R_V \equiv n^*_V = \lfloor (L + p_0)/(p_P + d_B) \rfloor \tag{19.25}$$

$$T^R_p \equiv T^*_p = \frac{1}{(L + p_P + p_0)} \lfloor \frac{L + p_0}{p_P + d_B} \rfloor \tag{19.26}$$

and the MIT is given by, according to Equation (19.22) and D^R_{k+1},

$$MIT^R \equiv MIT^* = \frac{(n^R_V - 1)n^R_V(p_P + d_B)}{2n^*_J} \tag{19.27}$$

19.3.2 Order Rule

The vehicle-routing rule is referred to as the *order rule*, if it deterministically and repeatedly allocates vehicles (and jobs) to stations, $S_{n_S}, S_{n_S-1}, \ldots, S_1$ in this order [222](see Fig. 19.1 and Fig. 19.2). Let

$$m(i) = n_S + 1 - i + \lfloor (i - 1)/n_S \rfloor n_S, i = 1, 2, \ldots, n_J \tag{19.28}$$

then, job J_i is served on station $S_{m(i)}$ by $V_{k(i)}$. Figure 19.4 illustrates the order rule with $n_S = 6$, $n_V = 4$, and $n_J = 12$ where the number in a parenthesis on each vehicle stands for the job served by that vehicle. In the 1st lap the 4 vehicles, respectively, serve on stations S_6 to S_3 without interference. In the 2nd lap V_1 and V_2, respectively, serve on stations S_2 and S_1 to keep the load balance on the processing stations. As a result, V_3 and V_4 are blocked behind S_1 and the 4 vehicles are divided into two subgroups;$\{V_1, V_2\}$ and $\{V_3, V_4\}$ (see

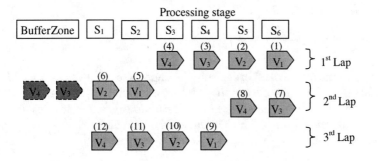

Fig. 19.4. Order rule with $n_S = 6$, $n_V = 4$ and $n_J = 12$

Fig. 19.3). In the 3rd lap the vehicles serve on S_4 to S_1 to keep the load balance. An outstanding feature of the order rule is that no interference occurs in any station but S_1 and I/O under constant processing times (Equation (19.2)). Let $g \equiv \gcd(n_S, n_V)$ be the greatest common divisor of n_S and n_V, then the order rule divides the set of n_V vehicles into $a \equiv n_{sg} = n_V/g$ subgroups $G_1, G_2, \ldots, G_a$ under assumptions (19.2) and (19.3) so that each subgroup has $n_{sw} = g$ vehicles in Fig. 19.3. Let $V_{p,q}$ be the qth vehicle of the pth subgroup $(p = 1, 2, \ldots, a, q = 1, 2, \ldots, g)$, then, MIF distance (19.9) from $V_{p,q+1}$ to $V_{p,q}$ and that from Equation (19.8) from $V_{p+1,1}$ to $V_{p,g}$ are, respectively, given by

$$D^O_{p,q+1} \equiv D^*_{p,q+1} = p_0 + d_B, \quad D^O_{p+1,1} \equiv D^*_{p+1,1} = p_P + d_B$$

thus, MFL (19.12) becomes

$$L^O_V(n_V, g) \equiv L^*_V = \sum_{p=1}^{a-1}\left(\sum_{q=1}^{g-1} D^O_{p,q+1} + D^O_{p+1,1}\right) + \sum_{q=1}^{g-1} D^O_{a,q+1}$$

$$= \frac{n_V(g-1)p_0 + (n_V - g)p_P + (n_V - 1)gd_B}{g} \qquad (19.29)$$

and the condition of the steady state equation (19.15) is satisfied, if

$$L \geq L^O_V(n_V, g) + p_P + d_B - p_0$$

$$= n_V\{(1 - \frac{1}{g})p_0 + \frac{1}{g}p_P + d_B\} - p_0 \qquad (19.30)$$

The throughput (19.17) leads to

$$T^O_p(n_V, g) \equiv T^V_p = \frac{1}{(L + p_P + p_0)} \lfloor \frac{L + p_0}{(1 - 1/g)p_0 + p_P/g + d_B} \rfloor \qquad (19.31)$$

if Equation (19.30) is satisfied. The mean interference time (19.22) satisfying Equation (19.30) is given by, according to $D^O_{p,q+1}$ and $D^O_{p+1,1}$,

$$MIT^O(n_V, g) \equiv MIT^*$$
$$= \frac{(n_V/2)[(p_P + d_B)(n_V/g - 1) + p_0 n_V(1 - 1/g)]}{n_J^*} \quad (19.32)$$

These results mean that the order rule depends on $g=\gcd(n_s, n_V)$ and hence the throughput does not always linearly increase with n_V. T_p^* and MIT^* of the order rule are better than those of the random rule except for $g = 1$ where both have the same values. This dependence of the order rule is overcome in new vehicle-routing rules as introduced in the next section.

19.4 Optimal Vehicle Rules

This section shows two vehicle routing rules: one routing rule (referred to as *exchange-order* and denoted E-order) is optimal as long as a steady state is realized, but not optimal when the OLB interference is inevitable. The other is the dynamic-order rule (denoted D-order) which is same as the E-order rule under the steady state and better under the OLB interference.

19.4.1 Exchange-Order Rule

Let the order rule be applied to the first $(l - 1)(l = 1, 2, \ldots)$ laps and the last vehicle, V_{n_V} serve job $J_{(l-1)n_V}$ on station S_{u+1} , then u is derived from Equations (19.1) and (19.28).

$$u = \{\lfloor \frac{(l-1)n_V - 1}{n_S} \rfloor + 1\}n_S - (l-1)n_V \quad (19.33)$$

In the lth lap V_1 serves on S_u (S_{n_V}, if $u = 0$) and V_{n_V} to S_w, where

$$w = \{\lfloor \frac{(ln_V - 1)}{n_S} \rfloor + 1\}n_S + 1 - ln_V \quad (19.34)$$

One of the following three exclusive conditions to keep the load balance holds. 1) $u < w - 1$: the number of vehicles allocated to stations $S_m(m = u + 1, u + 2, \ldots, w - 1$ is less than the one to each of the remaining stations by one. 2) $u = w - 1$: each station has the same number of vehicles allocated (when n_V is a multiple of n_S) 3) $u > w - 1$: the number of vehicles allocated to stations $S_m(m = w, w + 1, \ldots, u)$ is larger than the one to each of the remaining stations by one.

The E-order rule uses the following two subrules in the lth lap and keeps the load balance as the order rule does, although it may take a different routing rule (the E-order rule is, of course, applied from scratch ($l = 1$) in the execution):   **Changing rule:** the order rule is applied to a subset of n_V vehicles from station S_{n_S}. The changing rule is optimal for the specified subset, because it gives no interference except unavoidable ones (*e.g.* if $n_V > n_S$,

$(n_V - n_S)$ vehicles are blocked behind S_1 in any rule). **Unchanging rule:** the order rule is applied to a subset of n_V vehicles from S_u. The unchanging rule is optimal, if u is not smaller than the number of the vehicles in the specified subset, because it gives no interference and is the changing rule, if $u = n_S$.

The E-order rule uses these two rules depending on the sign of $(u - v)$ as follows. 4) $u \le w(u < n_V, n_S)$: The changing rule is applied to the first $(n_V - u)$ vehicles, $V_1, V_2, \ldots, V_{n_V - u}$ and then the unchanging rule is applied to the last u vehicles, $V_{n_V - u + 1}, \ldots, V_{n_V}$. It is easily seen that $V_{n_V - u}$ serves on S_w and $V_{n_V - u + 1}$ on S_u, thus the above conditions 1) and 2) are satisfied. 5) $u > w(u < n_V, n_S)$: The changing rule is applied to the first $(n_V - u + w - 1)$ vehicles and then the unchanging rule is applied to the last $(u - w + 1)$ vehicles, $V_{n_V - u + w}, \ldots, V_{n_V}$.

It holds by Equations (19.33) and (19.34) that

$$n_v - u + w - 1 = \{\lfloor (l^{n_V} - 1)/n_S \rfloor - \lfloor ((l - 1)n_V - 1)/n_S \rfloor\}n_S$$

is a multiple of n_S, *i.e.* each station serves the same number of jobs by the first $(n_V - u + w - 1)$ vehicles and the routing of the last $(u - w + 1)$ vehicles that starts from S_u satisfies the above condition 3), thus the E-order rule keeps the load balance as the order rule does. The detail of the E-order rule is as follows: Figure 19.5 illustrates the E-order rule with the same condition as in

---

Algorithm 19.1 E-order

1: $l \leftarrow 0, \ell_{\max} \leftarrow \lceil n_J / n_V \rceil$
2: $l \leftarrow l + 1$. If $l > l_{\max}$, halt. If $l = l_{\max}$ and $n_J < l_{max}n_V$, then $n_V \leftarrow n_J - (l_{\max} - 1)n_V$, and go to Step 3.
3: Compute u by Equation (19.33).
4: If $u \ge n_V$, or if $u = n_S$, apply the unchanging rule to all n_V vehicles, and go to Step 2, otherwise, go to Step 5.
5: Compute w by Equation (19.34). If $u < w$, go to Step 6, otherwise, go to Step 7.
6: Apply the changing rule to the first $(n_V - u)$ vehicles, and apply the unchanging rule to the last u vehicles, $V_{n_V - u + 1}, \ldots, V_{n_V}$. Go to Step 2.
7: Apply the changing rule to the first $(n_V - u + w - 1)$ vehicles, $V_1, \ldots, V_{n_V - u + w - 1}$, and the unchanging rule to the last $(u - w + 1)$ vehicles, $V_{n_V - u + w}, \ldots, V_{n_V}$. Go to Step 2.

Fig. 19.4 to compare with the order rule. In the 1st lap the vehicles take the same routing as the order rule by Step 4 ($u = 6 = n_S$). In the 2nd lap V_1 and V_2 serve on S_6 and S_5, and V_3 and V_4 on S_2 and S_1, respectively by Step 6 ($u = 2 < n_S$ in Step 4 and $u < w = 5$ in Step 5). This routing is different from the one by the order rule in Fig. 19.4, but keeps the same load balance as the order rule does. As a result, no interference occurs. In the 3rd lap the same routing as the one by the order rule is taken by Step 4 ($u = 4 = n_V$ in

Step 4). It is easily seen that the E-order rule separates the n_V vehicles into

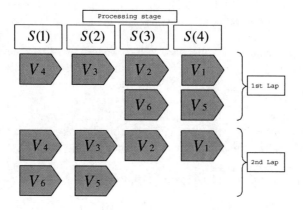

Fig. 19.5. E-order rule with $n_S = 6$, $n_V = 4$ and $n_J = 12$

$\lceil n_V / n_S \rceil$ subgroups, each of the first $\lfloor n_V / n_S \rfloor$ groups consists of n_S vehicles and the last one has $(n_V - \lfloor n_V / n_S \rfloor n_S)$ vehicles in a steady state. Thus, MFL (19.12) is given by

$$L_V^E(n_V) \equiv L_V^* = (p_P + d_B)(\lceil n_V / n_S \rceil - 1)$$
$$+ (p_0 + d_B)\{(n_S - 1)\lfloor n_V / n_S \rfloor + max(0, n_V - \lfloor n_V / n_S \rfloor n_S - 1)\} \quad (19.35)$$

Note that the max part of the right-hand side of the equation becomes 0, if n_V is a multiple of n_s, and becomes 1 otherwise. The steady state is realized by Equation (19.15), if

$$L \geq L_V^E(n_V) + p_P + d_B - p_0$$
$$= q p_P + \{max(q n_S, n_V - 1) - q\}p_0 + max(q n_S, n_V - 1)d_B \quad (19.36)$$

where $q = \lceil n_V / n_S \rceil - 1$. It is easily seen that L_V^E does not depend on $g = \gcd(n_V, n_S)$ and by Equations (19.29) and (19.35) that, $L_V^E \leq L_V^0$. This means by Equations (19.17), (19.18) and (19.29) that

$$T_p^E(n_V) \geq T_p^0(n_V, g) \quad (19.37)$$

holds under the steady state. MIT (19.22) for the E-order rule is given by, according to (19.35),

$$MIT^E(n_V) \equiv MIT^* = \frac{q\{n_V - (q+1)n_S/2\}p_P + (n_V - q)p_0}{n_J} \quad (19.38)$$

It is also easily seen by Equation (19.32) that

$$MIT^E(n_V) \leq MIT^0(n_V, g) \quad (19.39)$$

holds under the steady state.

19.4.2 Dynamic Order Rule

The E-order rule is not optimal when the OLB interference occurs. For example, consider when V_1 at the 2nd lap is blocked by V_4 at the 1st lap in Fig. 19.5, then, the 4 vehicles are blocked, because V_1 and V_2 are scheduled to serve on S_6 and S_5, respectively and hence cannot serve on S_2 and S_1, though they stay there, while the order rule can take a more expedient routing as shown in Fig. 19.4. The dynamic-order routing rule (denoted D-order routing) is obtained by modifying the E-order rule every time the OLB interference occurs. Let

$$D_{OLB} = L - L_V - p_0 \qquad (19.40)$$

where L_V stands for the fleet length of the n_V vehicles. Then, an OLB interference occurs, if $D_{OLB} < D^*_{n_V+1} = p_P + d_B$ as shown in Equation (19.15). The D-order rule is as follows:

Algorithm 19.2 D-order

1: the same as Step 1 of algorithm E-order.
2: the same as Step 2 of the E-order.
3: Compute u by Equation (19.33) and D_{OLB} by Equation (19.40).
4: If $u > n_V, u = n_S$, or $D_{OLB} < p_P + d_B$, apply the unchanging rule to all n_V vehicles, and go to Step 2, otherwise, go to Step 5.
5: the same as Step 5 of algorithm E-order.
6: the same as Step 6 of algorithm E-order.
7: the same as Step 7 of algorithm E-order.

It is obvious that algorithm D-order keeps the load balance as the order and the E-order rules do and is the same as algorithm E-order except Step 4 that allocates to idle stations as many vehicles as possible when the OLB interference occurs, resulting in not smaller throughput (and not larger interference time) than the E-order rule.

19.5 Numerical Simulation

This section describes results obtained by numerical simulation that was executed to confirm the theoretical results obtained so far. Figure 19.6 shows the effect of n_V (the number of vehicles) on T_P (the throughput) for each of the random, the order, the E-order and the D-order rules under conditions: $n_J = 600$ (the number of jobs), $n_S = 6$ (the number of processing stations), $p_P = 55$s (constant processing time of the processing stations), $p_0 = 1$s (constant processing time of the I/O station), $d_B = 5$m (the minimum distance between vehicles), $v_{max} = 1$m/s (the max. speed of the vehicle) and $L = 210$m

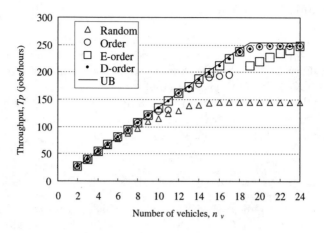

Fig. 19.6. Effect of n_V on T_P ($n_S = 6$, $n_J = 600$, $p_P = 55(s)$, $p_0 = 1(s)$, $L = 210(m)$)

(the loop length). The real line stands for upper bound (19.20) (consisting of two lines T_P^V (19.17) and T_p^S (19.19), respectively). The random rule saturates with lower T_P due to the early OLB interference. This fact is consistent with Equations (19.24) and (19.25). The order rule increases the throughput with n_Vs, but makes bumps with some n_Vs, depending on $\gcd(n_S, n_V)$ as discussed in Section 19.3.2. The E-order rule linearly increases T_P according to T_p^V till $n_V = 18$ and makes a big bump with $n_V \geq 19$, resulting in being behind the order rule as suggested in 19.4.2. The D-order rule keeps the same T_P as the E-order rule till $n_V = 18$ and the same T_P as the order rule with $n_V \geq 19$, resulting in keeping T_P close to the upper bound (19.20). Figure 19.7 shows the effect of n_V on the mean interference time $MIT(s)$ under the same condition as the one in Fig. 19.6 . For example compare the order rule and the E-order rule with $n_V = 10$ (and $\gcd(6,10)=2$) in Fig. 19.7. The MIF distance within a subgroup is, according to Equation (19.8).

$$d_B + p_0 v_{\max} = 5 + 1 = 6\text{m} \tag{19.41}$$

and the one between two adjacent subgroups is, according to Equation (19.9)

$$d_B + p_P v_{\max} = 5 + 55 = 60\text{m} \tag{19.42}$$

Then, in the order rule $L_V^0(10, 2) = 270$ (by Equation (19.29)), thus, the OLB interference is estimated to occur by Equation (19.30). $MIT^0(10, 2) = 2.0$ by Equation (19.32), but the real MIT is much larger due to the OLB interference. In the E-order rule $L^E(10) = 108$ (by Equation (19.35)) and hence the steady state is estimated to be realized by Equation (19.36). $MIT^E(10) = 0.38(s)$ (by Equation (19.38)) is close to the real value.

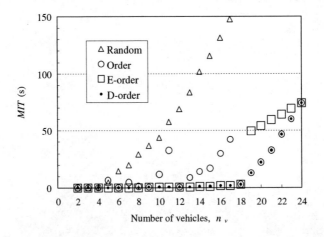

Fig. 19.7. Effect of n_V on mean interference time MIT with the same condition as Fig. 19.6

19.6 Concluding Remarks

Our study was motivated by the PCVRS in a real automated warehouse where complicated interference between vehicles was observed. The warehouse told us that there is a limit of increasing the number of vehicles to improve the throughput: a fleet size larger than this limit does not improve further or occasionally deteriorates the throughput. The reason had not been known and been suspected due to a vendor-provided software which is of a black-box type. We hope that our result is informative to the warehouse.

Tabu-based Optimization for Input/Output Scheduling

20.1 Introduction

We discuss an optimal scheduling problem for a large-scale automated warehouse (referred to as AS/RS) with many (high bay pallet) racks for storage and automated S/R (storage and retrieval) machines such as stacker cranes (see Fig. 3.3). The racks are set in parallel and S/R machines run in aisles between the racks to store and retrieve items simultaneously. Each S/R machine can move a few items at a time; larger numbers of items need to be partitioned into more than one subgroups to assign S/R machines so that each S/R machine by routing a rack stores the items of a subgroup into designated locations and retrieves the items of a subgroup stored. We are therefore required to make simultaneous decisions of partition, pairing of the items and routings of S/R machines so as to minimize the maximum tour time.

We discuss the computational complexity of the problem that includes the size of the solution space, and propose an approximation algorithm based on a tabu search and a heuristic. We demonstrate performances of the proposed algorithm and the S/R machine by means of numerical experiment based on practical conditions in real automated warehouses. This chapter is based mainly on the results of Kise and coworkers [223], since at this time we are not aware of other research dealing directly with this subject.

20.2 Optimal Input/Output Scheduling Problem

The AS/RS we consider consists of $n + 1$ parallel racks and n identical S/R machines. Each rack consists of a columns and b rows composing $c = ab$ cells, each having width w and height h. An S/R machine runs in both horizontal and vertical directions simultaneously. The horizontal run is subject to the maximum speed V_{Xmax}, acceleration α_X and deceleration β_X, and the vertical run limited to the maximum speed V_{Ymax}, acceleration α_Y and deceleration β_Y. The time the S/R machine takes to travel between two cells is measured by the max. of the horizontal and vertical times (*i.e.* $L\infty-$ norm). The capacity of an S/R machine, m is defined by the number of arms it has, each being able to hold one item, so that the S/R machine can hold m items at a time. The tour starts with items to store and finishes with items retrieved at the input/output location of the rack. Tour time is defined by the time units a tour takes.

Our AS/RS adopts a policy that every rack should have the same function such that any item can be stored and retrieved at any rack for maximum reliability and efficiency. It is assumed without loss of generality that there are M items to input (*i.e.* store) and M items to output (*i.e.* retrieve) and $M = nm$; otherwise dummy items located at the input/output location are added. Let I be the set of M input items and O the set of M output items. They are partitioned into n input subgroups, $I_1, I_2, \ldots, I_n$ and n output subgroups, $O_1, O_2, \ldots, O_n$, each having m items. Then, a pair of one input and one output subgroups is associated to an S/R machine. A routing of an S/R machine (I/O routing) is defined by a sequence of visiting $2m$ cells, the locations of which are designated in advance. The tour time depends on partitions of input and output (denoted I/O) items, I/O pairs and I/O routes to be performed. Thus a scheduling problem (referred to as the optimal I/O scheduling problem) for an AS/RS is to simultaneously decide I/O partitions, I/O pairs and I/O routes so as to minimize the maximum tour time over n S/R machines.

Many scheduling models related to AS/RSs have received extensive attention in the literature (see *e.g.* [224] for a survey). However, most of them have concentrated on sequencing (*i.e.* touring) for a single S/R machine to visit many cells in a rack or racks. Those implicitly assume a man-aboard S/R machine that can serve many items in a single tour. One of the most fundamental models is the "order picking" problem in which only retrieving operations are performed (see *e.g.* [225] again for a survey). On the other hand our model assumes unmanned S/R machines that cannot serve many items in a single tour due to technological constraints.

20.3 Computational Complexity

The solution space of the optimal I/O scheduling problem can be obtained by the product of the numbers of I/O partitions, the I/O pairs and the I/O routes. It is easily seen that the number of different input (and output) partitions is given by

$$g(M, m) = \frac{M!}{(m!)^n n!} \tag{20.1}$$

and the number of different I/O pairs by

$$p(M, m) = \frac{g^2(M, m)}{n!} = \frac{(M!)^2}{(m!)^{2n} n!} \tag{20.2}$$

Any feasible tour has to keep the SC capacity constraint to hold at most m items, *i.e.* at any time in a tour the number of items stored so far must be equal to or greater than the number of items retrieved. As a result the number of feasible tours is given by

$$r(m) = \frac{(2m)!}{(m+1)} \tag{20.3}$$

It is interesting to note that this number is equivalent to the Catalan number that is typically the number of combinations. Thus, the size of solution space results in

$$s(M, m) = r(m)p(M, m) = \frac{(2m)!(M!)^2}{(m+1)n!(m!)^{2n}} \tag{20.4}$$

Let $P(M, m)$ be the optimal I/O scheduling problem with M items and m arms. A *NP*-hard 3-dimensional matching problem (see *e.g.* [40] for the *NP*-hardness) can be reduced to a special case of $P(M, 2)$ in which each tour consists of one input item (with a dummy input one) and two output items. $P(M, M)$ that is an order picking problem is generally *NP*-hard, although there are some solvable cases in specially structured AS/RSs (see e.g. [225] [226]).

It is concluded that the optimal I/O scheduling problem is generally *NP*-hard, and hence an efficient approximation algorithm is needed in practice. We propose an approximation algorithm based on tabu search and a heuristic in the next section.

20.4 Approximation Algorithm

As discussed in the preceding sections, the solution consists of the I/O partition, the I/O pair and the I/O route. Our algorithm uses a simple tabu search for the I/O partitioning and a heuristic for the I/O pairing and the I/O routing. Here we show only components composing the tabu search, which is introduced in Chapter 20 (see, *e.g.* [227] for a more comprehensive survey of the tabu search).

Solution in our tabu search is represented by a list of input subgroups, *i.e.* $I = (I_1, I_2, \ldots, I_n)$ where $I_k = (i_{k1}, i_{k2}, \ldots, i_{km})$, $k = 1, 2, \ldots, n$ stands for a list of m input items to be visited by the kth S/R machine. An initial list is given by a random sequence (the entire I/O schedule is computed based on I using a heuristic, as shown later).

Objective function is of bicriterion, *i.e.* let $tt(I_k)$ be the tour time of the kth subgroup, and then we adopt the max. tour time and the mean tour time as objective functions that are, respectively, defined by

$$TT_{\max}(I) = \max\{tt(I_1), tt(I_2), \ldots, tt(I_n)\} \tag{20.5}$$

$$TT_{mean}(I) = \frac{tt(I_1) + tt(I_2) + \ldots + tt(I_n)}{n} \tag{20.6}$$

The first priority is given to the max. tour time and the second one to the mean tour time.

Neighborhood: Given solution I, its neighborhood solution $I' = (I_1', I_2', \ldots, I_n')$ is obtained by the following procedures: Let I_k be a subgroup with the max tour time, $i.e.$ $TT_{\max}(I) = tt(I_k)$, breaking tie by the max. mean tour time. Then, I_k has item i_{kj}, $j = 1, 2, \ldots, m$, to move for I' as follows:

a) i_{kj} $(j = 2, 3, \ldots, m)$ moves to the first position in I_k.

b) i_{kj} $(j = 1, 2, \ldots, m)$ and every element in other subgroups is exchanged. The size of the neighborhood $N(I)$ results in

$$| N(I) | = m - 1 + m(M - m) \tag{20.7}$$

Procedures a) and b) change at most two subgroups in the neighborhood, and its size is smaller than $M(M - 1)/2 = mm(mn-1)/2$, the number of solutions obtained by exchanging all pairs in M items. The search is thus effective for getting better neighborhood solutions efficiently by using a good heuristic for pairing and touring, as discussed later.

Tabu list is given by list $TL = (TL(1), TL(2), \ldots, TL(M))$ where $TL(i)$ stands for the number of iterations in the tabu search during which item i is prohibited to move, $i.e.$ only item i with $TL(i) = 0$ can move in the next iteration. Each $TL(i)$ is given by an integer random number in the range $[TL_{min}, TL_{\max}]$ and $TL(j) = \max\{TL(j) - 1, 0\}$, $j \neq i$ each time i moves. TL_{min} and $TL_{\max}$ are parameters given by the user. This strategy aims at diversity of the tabu search.

Move function $mv(I) = I'$ is defined by I' with the minimum $TT_{\max}(I')$, breaking tie by the minimum $TT_{mean}(I')$ among neighborhood solutions that are not prohibited by the tabu list.

Terminal condition: The search stops when it reaches $t_{\max}$ iterations. $t_{\max}$ is a parameter given by the user.

A schedule (including the I/O pair and the I/O route) is obtained from solution $I = (I_1, I_2, \ldots, I_n)$ by a modification of the well-known nearest neighbor rule [5] (denoted P/R heuristic).

P/R heuristic: The kth $(k = 1, 2, \ldots, n)$ tour with m input items first takes the first element i_{k1} of I_k and then the nearest cell (measured by time) within $\{I_k - \{i_{k1}\}\} \cup O_k$, where O_k stands for the set of output items that are not delivered so far. In general, assume that the tour already served p input items and q output items, then if $p > q$, serve a nearest cell from among input and output items that have not been served yet, otherwise store the one from among the remaining input items to hold the capacity constraint of the S/R machine.

A concluding remark of this section is that the search space is very sparse compared with the solution space shown in Equation (20.4). We therefore perform some numerical experiments to demonstrate the effect of the approximation algorithm proposed.

20.5 Numerical Experiment

We demonstrate performances of the approximation algorithm proposed and performance of the S/R machine by means of a numerical experiment based on practical conditions in real AS/RSs.

Experimental conditions are as follows.

1. Rack condition: a=20, b=50, $c = ab = 1000$, $w = h$=1.5 m and n=10 are fixed (see Section 20.2).
2. S/R machine condition: V_{Xmax}=1.67 m/s and V_{Ymax}=0.5 m/s are fixed. $\alpha_X = \beta_X = \alpha_Y = \beta_Y = 0.3$, 0.4 and 0.5 m/s^2 are tested. Times for storage and retrieval motions are neglected since they do not affect the scheduling. Four capacities, m=1,2,3 and 4 are tested. Note that AS/RS with $m > 2$ is rare in real AS/RSs, but m=3, 4 are technologically possible.
3. Item condition: M=60 is fixed. The cell locations for items to be served are randomly generated in range [1, c] without duplication. The results shown here exclude tours with dummy items (in the case of m=4).
4. Tabu list parameters: TL_{min}=7 and TL_{max}=20+0.1M are fixed .
5. Terminal condition: Several numbers of t_{max} from 1 to 500 are tested.
6. The program is coded in C and executed on a PC (Pentium 4, 2.4 GHz and 512MB RAM).

The following is the result with $\alpha_X = \beta_X = \alpha_Y = \beta_Y = 0.3$, $m = 3$ and $t_{max} = 300$ unless otherwise indicated.

The result on performance of the approximation algorithm is as follows. Table 20.1 shows the effect of the max. number of iterations, t_{max} on max. tour time, TT_{max} s, mean tour time, TT_{mean} s and computation time, CPU s. t_{max}=0 and t_{max}=1 mean initial solution and one of the single local search, respectively. The single local search and the tabu search with t_{max}=300, respectively, reduce TT_{max} by about 18% and about 25% from the initial solution by spending 0.07 s and 223 s CPU times. TT_{mean} is also reduced.

Performance of S/R machine is shown in Table 20.2 and Table 20.3. TP in Table 20.3 stands for the (average) throughput (rate) that is the number of items processed per hour. Obviously minimizing TT_{max} is equivalent to maximizing TP. Increasing m is very effective for making AS/RS efficient. Table 20.3 shows the effect of acceleration (and deceleration) on the max. tour time. By increasing from 0.3 to 0.5 m/s^2 the max. tour time is reduced by about 5%. It is interesting to compare this effect with the one by the optimizing scheduling in Table 20.1.

20.6 Concluding Remarks

This chapter described an optimal input and output scheduling problem for an AS/RS and proposed a tabu-based approximation algorithm. It was demonstrated to be effective in real AS/RSs by numerical experiment. There remain

Table 20.1. Effect of max. no. of iterations, t_{max} on TT_{max}, TT_{mean} and CPU

t_{max}	0	1	10	50	100	200	300	400	500
TT_{max}	183.8	151.5	144.6	141.9	141.0	139.8	138.9	138.8	137.8
TT_{mean}	119.2	116.8	116.2	115.6	115.2	115.1	115.0	115.3	115.0
CPU	< 0.01	0.74	7.40	37.2	74.4	149	223	298	372

Table 20.2. Effect of capacity of S/R machine, m on max. tour time, TT_{max} and throughput, TP

m	1	2	3	4
TT_{max} | TP	123.5 | 583.0	130.1 | 1107	138.9 | 1554	149.1 | 1932

Table 20.3. Effect of acceleration/deceleration ($\alpha = \beta$) on max. tour time, TT_{max}

$\alpha = \beta$	0.3	0.4	0.5
TT_{max}	138.9	134.5	132.7

some issues to be studied for efficient AS/RSs. In many AS/RSs there exist automated material handling systems (MHS) which are introduced in Chapter 19. Then, a generalized optimal scheduling between the MHS and S/R machines will therefore be effective for efficient AS/RS.

Program Listings

A Reference Implementation of the S-ring

The C program "sring.c":

```c
/*
 * sring.c - reference implementation of an S-ring
 * (C) 1999 - 2006,  S. Markon, Osaka, Japan. Released under the GPL
 * This is a fragment, for complete source see the book's web site.
 */

#define N 32
static int s[N], c[N], p, pa, r, n, m, nw, it;
static double w[N+N],l;

/* Utility: return TRUE with probability p */
static int prob(double p) {
  return ((double)(rand()) / (double)RAND_MAX) < p;
}

/* Initialize parameters and work variables */
static void init_ring(int nx, int mx, double lx, int sd) {
    int i;

    srand(sd);               /* seed from user */
    n = nx;                  /* number of sites */
    m = mx;                  /* number of servers */
    l = lx;                  /* probability of arrivals */
    r = n;                   /* free sites initialized */
    for (i=0;i<n;i++) {
      s[i] = c[i] = 0;
      if (prob(l)) {
        c[i] = 1;            /* generate customers */
        r--;                 /* and decrease free count */
```

```
      }
    }
    for (i=0;i<m;i++) {
      s[i] = 1;                      /* generate m servers */
    }
    p = 0;                           /* current site position */
    pa = 1;                          /* next site for server */
  }


  /* Step the ring from s to s1; a: decision  d: customer arrival */
  static int step_ring(int a, int d) {
    if (d && !c[p]) {                /* new customer? */
      c[p] = 1; r--;                 /* decrease free site count */
    }
    if (s[p]) {                      /* a server here? */
      if (c[p]) {                    /* and also a customer? */
        if (a || s[pa]) {            /* and to be served? */
          c[p] = 0; r++;             /*   serve; increase free sites */
        } else {                     /* or not to be served? */
          s[pa] = s[p]; s[p] = 0;    /*   pass by */
        }
      } else if (!s[pa]) {           /* server free to proceed? */
        s[pa] = s[p]; s[p] = 0;      /*   go ahead */
      }
    }
    pa = p; p = (p + n - 1)%n;       /* move current-site-position */
    return r;                        /* number of free sites now */
  }


  /* Calculate decision by a single-layer perceptron */
  static int perceptron(int n, double *w) {
    int i,j;
    double x=0.0;

    for (i=0,j=p;i<n;i++,j=(j+1)%n)
      x += (double)s[j]*w[i]         /* contribution of servers */
         + (double)c[j]*w[i+n];      /* contribution of customers */
    return x>0.0;                    /* Heaviside function */
  }
```

A simple shell script "sring.sh" for exercising "sring.c":

```
#!/bin/sh
echo -n "greedy policy: "
./sring n6 m2 i1000000 p0.3 r0$1  1 1 1 1 1 1  1 1 1 1 1 1
echo -n "balance policy: "
./sring n6 m2 i1000000 p0.3 r0$1 -1 3 3 3 3 -1 0 0 0 0 0 0
echo -n "optimal policy: "
./sring n6 m2 i1000000 p0.3 r0$1 -15 33 33 33 21 -3 0 -8 -4 -4 12 4
```

An Implementation of the Simple Stochastic Search

The C program "sss.c":

```c
#include <stdio.h>
#include <stdlib.h>
#include <math.h>
#include <float.h>
#include <unistd.h>
#define  DEBUG 0
#define Z (rand()/(RAND_MAX+1.0))
/* Utility: return TRUE with probability p */
static int prob(double p)
{
  return ((double)(rand()) / (double)RAND_MAX) < p;
}
double uni(double a, double b)
{
  return a + (b-a) * (drand48()) ;
}

void main(int argc, char *argv[])
{
  double e=1.0; /* step size */
  double d=0.5; /* distance delta */
  double p=0.4; /* probability of generating a better candidate */
  double tau=0.811; /* threshold value */
  int steps, sim, l;
  double i, j, tmp_i, result;
  int seed = 13232; /*random seed */
  for (l=1;l<argc;l++) {
    switch ( argv[l][0] ) {
    case 'd': d = atof(argv[l]+1); break;
    case 'p': p = atof(argv[l]+1); break;
    case 't': tau = atof(argv[l]+1); break;
    case 's': seed = atoi(argv[l]+1); break;
    }
  }
  srand48(seed);
  for (sim=1; sim <= 1000; sim++){
      i = 0.0; tmp_i= 0.0;
      for (steps = 0; steps < 1; steps++){
          if (drand48() < p){
            j=(i+1)*d+uni(-e/2.0,e/2.0);
              tmp_i=i*d+uni(-e/2.0,e/2.0);
            if (j  > tmp_i + tau) i++;}
          else{
            j=(i-1)*d+uni(-e/2.0,e/2.0);
              tmp_i=i*d+uni(-e/2.0,e/2.0);
```

```
              if (j  > tmp_i + tau) i--;
            }
            steps++;
         } /* end steps */
       result = result + i;} /* end sim */
    result /=sim;
  }
```

References

1. V.M. Lampugnani, L. Hartwig, and J. Simmen (eds.). *Lift, Elevator, Paternoster — A Cultural History of Vertical Transport*. Vertical. Ernst, Berlin, 1994.
2. G.R. Strakosch. *Vertical Transportation: Elevators and Escalators*. John Wiley & Sons, New York, 1967.
3. G.R. Strakosch (ed.). *The Vertical Transportation Handbook*. John Wiley & Sons, New York, 1998.
4. G. Barney. *Elevator Traffic Analysis, Design and Control*. Cambridge U.P., 1986.
5. G. Barney. *Elevator Traffic Handbook*. Spon Press, London, 2003.
6. A.T. So and Wai Lok Chan. *Intelligent Building Systems*. Kluwer International Series on Asian Studies in Computer and Information Science. Kluwer, Boston, 1999.
7. M.L. Siikonen and K. Suihkonen. People Flow and Automated Transportation with Hospital Elevators. *Business briefing: hospital engineering and facilities management, 2005, issue 2*, pages 41–44, 2005.
8. R.D. Peters. Ideal Lift Kinematics. In *Elevator Technology 6, Proc. of ELEVCON'95, ed. G.C. Barney*, pages 175–184, Stockport, 1995. IAEE.
9. S. Tanaka, Y. Uraguchi, and M. Araki. Dynamic optimization of the operation of single-car elevator systems with destination hall call registration: Part I. Formulation and simulations. *European Journal of Operational Research, Vol. 167*, pages 550–573, 2005.
10. S. Tanaka, Y. Uraguchi, and M. Araki. Dynamic optimization of the operation of single-car elevator systems with destination hall call registration: Part II. The solution algorithm. *European Journal of Operational Research, Vol. 167*, pages 574–587, 2005.
11. G.D. Closs. The Computer Control of Passenger Traffic in Large Lift Systems. *Ph.D. Thesis, U. Manchester*, 1970.
12. B.A. Powell. Elevator Planning and Analysis on the Web. In *Elevator Technology 11, Proceedings of ELEVCON'01*, pages 113–121. IAEE, 1998.
13. R.D. Peters. Simulation for control system design and traffic analysis. In *Elevator Technology 9, Proceedings of ELEVCON'98*, pages 226–235. IAEE, 1998.

14. N.T.J. Bayley. On Queuing Processes with Bulk Service. *J. Roy. Stat. Soc. B, Vol.16, No.1*, pages 80–87, 1954.

15. N.A. Alexandris, C.J. Harris, and G.C. Barney. Derivation of the Mean Highest Reversal Floor and Expected Number of Stops in Lift Systems. *Appl. Math. Modelling Vol.3, No.4*, pages 275–279, 1979.

16. N.A. Alexandris, G.C. Barney, and C.J. Harris. Multi-car Lift System Analysis and Design. *Appl.Math.Modelling, Vol.3, No.4*, pages 269–274, 1979.

17. N.A. Alexandris, C.J. Harris, and G.C. Barney. Evaluation of the Handling Capacity of Multi-Car Lift Systems. *Appl.Math.Modelling, Vol.5, February 1981*, pages 49–52, 1981.

18. K. Hirosawa. Numerical Solution of Bulk Queues by Using Jaiswall Model. *J.I.E.E.J. Vol.91, No.3*, pages 492–500, 1971.

19. K. Hirosawa, K. Kawatake, and T. Yuminaka. Elevator Service Evaluation Index and Its Measuring Device (in Japanese). *Hitachi Hyouron, Vol.53, No.6*, pages 6–12, 1971.

20. T. Hatanaka, T. Takine, Y. Takahashi, and T. Hasegawa. Modeling and Performance Evaluation of an Elevator System. *Trans. IEICE, Vol.6, No.9*, pages 396–406, 1993.

21. P.R. Tregenza. The Prediction of Passenger Lift Performance. *Architectural Science Review, September 1972*, pages 49–54, 1972.

22. R.D. Peters. Lift Traffic Analysis Formulae for the General Case. *Elevator World, December*, pages 30–31, 1990.

23. W.D. Kelton, R.P. Sadowski, and D.A. Sadowski. *Simulation with Arena*. Industrial Engineering. McGraw-Hill, Boston, MA, 1998.

24. J. Banks, II J.S. Carson, and B.L. Nelson. *Discrete-Event System Simulation (second ed.)*. Prentice Hall international series in industrial and systems engineering. Prentice Hall, New Jersey, 1996.

25. L.R. Al-Sharif and G.C. Barney. The Use of Moving Averages in Filtering Lift Traffic Patterns. *Control Systems Centre Report No.747, UMIST, Manchester, UK*, 1991.

26. L.R. Al-Sharif and G.C. Barney. The Inverse SP Method — Deriving Lift Traffic Patterns from Monitored Data. *Control Systems Centre Report No.745, UMIST*, 1991.

27. L.R. Al-Sharif. New Concepts in Lift Traffic Analysis — The Inverse SP Method. In *Elevator Technology 4, Proc. of ELEVCON'92, ed. G.C.Barney*, pages 8–17, Stockport, 1995. IAEE.

28. I. Mitrani. *Simulation Techniques for Discrete Event Systems*. Cambridge Computer Science Texts 14. Cambridge University Press, Cambridge, 1982.

29. J.S. Ostroff. *Temporal Logic for Real-Time Systems*. Advanced Software Development Series. Research Studies Press, Taunton, 1989.

30. C.A.R. Hoare. *Communicating Sequential Processes*. Prentice-Hall International Series in Computer Science. Prentice-Hall, Englewood Cliffs, 1985.

31. INMOS Limited. *occam 2 Reference Manual*. Prentice-Hall International Series in Computer Science. Prentice Hall, New York, 1988.

32. J. Hinton and A. Pinder. *Transputer Hardware and System Design*. Prentice Hall, New York, 1993.

33. J. Kramer, J. Magee, M. Sloman, K. Twidle, and N. Dulay. The Conic programming language: version 2.4. Technical report, Department of Computer Science, Imperial College of Science and Technology, 1984.

34. M. L. Siikonen. *Planning and Control Models for Elevators in High-Rise Buildings*. PhD thesis, Helsinki Unverstity of Technology, Systems Analysis Laboratory, Finland, October 1997.

35. V. T'kindet and J.C. Billaut. *Multicriteria Scheduling Theory Models and Algorithms*. Springer-Verlag, Berlin Heidelberg New York, 2002.

36. T.P. Bagchi. *Multiobjective Scheduling by Genetic Algorithms*. Kluwer Academic Pub., Boston Dordrecht London, 1999.

37. M. Pinedo. *Scheduling Theory, Algorithms, and Systems*. Prentice Hall, Englewood Clifs, New Jersey, 1995.

38. S.M. Johnson. Optimal two- and three stage production schedules with set-up time included. *Naval Research Logistics Quarterly, 1.*, pages 61–68, 1954.

39. P.C. Gilmore and R.E. Gomory. Sequencing a One-State Variable Machine: A Solvable Case of the Traveling Salesman Problem. *Operations Research 12*, pages 655–679, 1956.

40. M.R. Garey and D.S. Johnson. *Computer and Intractability — A Guide to the Theory of NP-Completeness*. W.H. Freeman and Company, San Francisco, 1986.

41. P. Brucker. *Scheduling Algorithms*. Springer, Berlin Heidelberg New York, 1995.

42. T. Ibaraki. Enumerative Approaches to Combinatorial Optimization, Part I. *Annals of Operations Research 10*, 1987.

43. T. Ibaraki. Enumerative Approaches to Combinatorial Optimization, Part II. *Annals of Operations Research 11*, 1987.

44. J. Cheng, H. Kise, and H. Matsumoto. A Branch-and-Bound Algorithm with Fuzzy Inference for a Permutation Flowshop Scheduling Problem. *European Journ. Of Operational Research, 96*, pages 578–590, 1997.

45. T.E. Morton and D.W. Pentico. *Heuristic Scheduling Systems with Applications to Production Systems and Project Management*. John Wiley & Sons, New York, 1993.

46. I.H. Osman and J.P. Kelly (eds). *Meta-Heuristics: Theory and Applications*. Kluwer Academic Pub., Boston Dordrecht London, 1996.

47. M.S. Fox. *Constraint-directed search: A case study in job shop scheduling*. Ph.D. Thesis, CMU=CS=RT-83-22, Carnegie Mellon University, Department of Computer Science, 1983.

48. M. Zweben and M.S. Fox. *Intelligent Scheduling*. Morgan Kaufmann, San Francisco, California, 1994.

49. E.M. Goldratt. *The Goal.* North River Press, 1984.

50. M.P. Groover. *Automation, Production Systems, and Computer-Integrated Manufacturing*. Prentice-Hall, Inc., Englewood Cliffs, New Jersey, 1987.

51. H. Inaba and S. Sakakibara. Flexible automation–unmanned machining and assembly cells with robots. In *Proc. of the 1st International Conf. on Flexible Manufacturing Systems*, 1994.

52. Y. Crama. Combinatorial Optimization Models for Production Scheduling in Automated Manufacturing Systems. *European Journ. Of Operational Research 99*, pages 136–153, 1997.

53. M. Dawande, H.N. Geismar, S.P. Sethi, and C. Sriskandarajah. Sequencing and Scheduling in Robotic Cells: Recent Developments. *Journ. of Scheduling, V.8*, pages 387–426, 2005.

54. D.E. Knuth. *The Art of Computer Programming, Vol.3*. Addison-Wesley, Reading, MA, 1973.

55. D. Levy. *Optimal Control in Vertical Transportation, Ph.D. Thesis.* Technion, Israel, 1973.

56. D. Levy, M. Yadin, and A. Alexandrovitz. Optimal Control of Elevators. *Int. J. Systems Sci. Vol.8, No.3*, pages 301–320, 1977.

57. S.O. Krumke. *Online Optimization - Competitive Analysis and Beyond.* Hab. Thesis, Technische Universitaet Berlin, Berlin, 2001.

58. M. Groetschel. *Online algorithms for transport systems.* Thesis, Technische Universitaet Berlin, Berlin, 1999.

59. H. Takeda. Latest Trends in Elevator Group Control Systems. *The Magazine of Building Equipment, No.464*, pages 51–55, 1989.

60. H. Aoki and K. Sasaki. Elevator Systems in Intelligent Buildings. *Proceedings of the Total Building Conference Singapore A2*, pages 50–62, 1990.

61. F. Ishikawa, T. Yamaguchi, and K. Haruki. Acquisition of Control Rule Knowledge for Elevator Group Controller. *Proc. 7th Sim.Techn.Conf.*, pages 93–96, 1988.

62. S. Kirkpatrick. Optimization by Simulated Annealing. *Science, Vol. 220*, pages 671–680, 1983.

63. G. Dueck and T. Scheuer. Threshold Accepting — A General Purpose Optimization Algorithm Superior to Simulated Annealing. *J. Comp. Phys., Vol. 90*, pages 161–175, 1990.

64. J.H. Holland. *Adaptation in Natural and Artificial Systems.* The University of Michigan Press, Ann Arbor, 1975.

65. S. Markon and Y. Nishikawa. TA Optimization for Elevator Group Control. *Proc. 34th Symp. of Control Eng.*, pages 351–354, 1990.

66. S. Lao, S. Markon, and Y. Nishikawa. The Genetic Algorithm as Applied to the Optimal Call Allocation for Elevator Group Control. *Proc. 35th Annual Conf. ISCIE*, pages 347–350, 1991.

67. N. Hikida, S. Tsuji, and K. Komaya. Service Limits of Elevators — Optimization by the S.A. Method. *Proc.1985 Annual Conf.J.I.E.E.*, pages 1931–1932, 1985.

68. S. Lao, S. Markon, and Y. Nishikawa. GA as Applied to Constructing an Elevator Group Control System. *Proc. 36th Annual Conf. ISCIE*, pages 85–86, 1992.

69. Y. Uesaka and K. Ozeki. *Pattern Recognition and Learning Algorithms.* Bunitsu Sogo Shuppan, 1990.

70. S. Haykin. *Neural Networks, A Comprehensive Foundation.* Macmillan, 1994.

71. R. Battiti. First- and second-order methods for learning between steepest descent and Newton's method. *Neural Computation*, 4:141–166, 1992.

72. H. Imano and H. Yamashita. *Nonlinear Programming.* Nikka Giren, 1978.

73. M. F. Moller. A Scaled Conjugate Gradient Algorithm for Fast Supervised Learning. *Neural Networks*, 6:525–533, 1993.

74. K. Saito and R. Nakano. Partial BFGS Update and Efficient Step-Length Calculation for Three-Layer Neural Networks. *Neural Computation*, 9:123–141, 1997.

75. H. Kita. Generalization Ability of Neural Networks. *Trans. ISCIE*, 36(10):625–633, 1992.

76. M. Kawato. *Computational Theory of the Brain.* Sangyo Tosho, 1996.

77. H. Nakanishi, T. Koda, and K. Inoue. Construction of an Optimal Feedback System with Neural Networks. *Trans. SICE*, 33:882–889, 1997.

78. I. Ono, M. Yamamura, and H. Kita. Real-valued GA and its Applications. *J.AIJ*, 15(2):259–266, 2000.

79. C.R. Reeves. *Modern Heuristics*. McGraw Hill, 1997.

80. K. Deb. *Multi-Objective Optimization using Evolutionary Algorithms*. Wiley, 2001.

81. W. Banzhaf. *Genetic Programming, An Introduction*. Morgan Kaufmann, 1998.

82. H. Kita and Y. Sano. Optimization of Uncertain Fitness Functions by GA. *J.AIJ*, 18(5):510–516, 2003.

83. D.E. Goldberg. *Genetic Algorithms in Search, Optimization, and Machine Learning*. Addison-Wesley, 1989.

84. N. Sannomiya, H. Kita, H. Tamaki, and T. Iwamoto. *Genetic Algorithms and Optimization*. Asakura, 1998.

85. H. Kitano. *Genetic Algorithms 4*. Sangyo Tosho, 2000.

86. Y. Nagata and S. Kobayashi. Crossover for the TSP — Proposal and Evaluation of the Branch Construction Crossover Method. *J.AIJ*, 14(5):848–859, 1999.

87. T. Yamada and R. Nakano. Solution of a Jobshop Scheduling Problem by Genetic Local Search. *Trans.JIPS*, 38(6):1126–1138, 1997.

88. K. Ikeda. *Global Optimization with GA and its Applications, PhD Thesis*. TIT, 2003.

89. S. Tsutsui and A. Ghosh. Genetic algorithms with a robust solution searching scheme. *IEEE Trans. Evolutionary Computation*, 1:201–208, 1997.

90. S. Kurahashi and T. Terano. Bayesian Optimization Methods and Estimation of Distribution Algorithms. *J.AIJ*, 18(5):487–494, 2003.

91. D.-Y. Cho and B.-T. Zhang. Continuous Estimation of Distribution Algorithms with Probabilistic Principal Component Analysis. In *Proc. of the 2001 IEEE Congress on Evolutionary Computation*, pages 521–526, 2001.

92. S.-Y. Shin and B.-T. Zhang. Bayesian Evolutionary Algorithms for Continuous Function Optimization. In *Proc. of the 2001 IEEE Congress on Evolutionary Computation*, pages 508–515, 2001.

93. J. Sakuma. New Developments in Real-valued GA based on the Estimation of Probability Distributions. *J.AIJ*, 18(5):479–486, 2003.

94. J. Branke. Creating Robust Solutions by Means of Evolutionary Algorithms. In *Proc. PPSN V*, pages 119–128, 1998.

95. H. Tamaki and T. Arai. A Genetic Algorithm Approach to Optimization Problems in an Uncertain Environment. In *Proc. of 1997 Int'l. Conf. on Neural Information Processing and Intelligent Information Systems*, pages 436–439, 1997.

96. K. Tanooka, H. Tamaki, S. Abe, and S. Kitamura. A Continuous Age Model of Genetic Algorithms Applicable to Optimization Problems with Uncertainties. In *Proc. IEEE SMC'99*, volume 1, pages 637–642, 1999.

97. D. Arnold and H.-G. Beyer. Local Performance of the the $(\mu/\mu_I, \lambda)$-ES in a Noisy Environment. In *Proc. FOGA 6*, 2000.

98. D. Arnold and H.-G. Beyer. Local Performance of the the $(\mu/\mu_I, \lambda)$-ES in a Noisy Environment. In *Proc. PPSN VI*, pages 39–48, 2000.

99. U. Hammel and T. Bäck. Evolution Strategies on Noisy Functions. How to Improve Convergence Properties. In *Proc. PPSN III*, pages 159–168, 1994.

100. H.-G. Beyer. Mutate Large, But Inherit Small! On the Analysis of Rescaled Mutations in $(1, \lambda)$-ES with Noisy Fitness Data. In *Proc. PPSN V*, pages 109–118, 1998.

101. S. Markon, D.V. Arnold, T. Bäck, T. Beielstein, and H.-G. Beyer. Thresholding – a selection operator for noisy ES. In *Proc. 2001 IEEE Congress of Evolutionary Computation*, pages 465–472, 2001.

102. S. Markon, D.V. Arnold, T. Bäck, T. Beielstein, and H.G. Beyer. Thresholding — a Selection Operator for Noisy ES. In *Proc. Congress on Evolutionary Computation (CEC'2001, Seoul)*, volume 1, pages 465–472. IEEE, 2001.

103. J.M. Fitzpatrick and J.J. Greffenstette. Genetic algorithms in noisy environments. *Machine Learning*, 3:101–120, 1998.

104. A. Aizawa and B.W. Wah. Scheduling of Genetic Algorithms in a Noisy Environment. *Evolutionary Computation*, pages 97–122, 1994.

105. Peter Stagge. Averaging efficiently in the presence of noise. In A. Eiben, editor, *Parallel Problem Solving from Nature, PPSN V*, pages 188–197, Berlin, Heidelberg, New York, 1998. Springer.

106. Y. Sano and H. Kita. Optimization of Noisy Fitness Functions by means of Genetic Algorithms using History of Search. *Trans. IEEJ*, 122-C(6):1001–1008, 2002.

107. Y. Sano and H. Kita. Optimization of Noisy Fitness Functions by means of Genetic Algorithms using History of Search. In *Proc. PPSN VI*, pages 571–580, 2000.

108. M. Yamamoto, T. Hashiyama, and S. Okuma. Simultaneous Computation of Controllers and Shapes in the Real World using Estimation of the Evaluation Function. In *10th Intelligent System Symposium*, 2000.

109. Y. Hanaki, T. Hashiyama, and S. Okuma. Reduction in the Computing Costs of GA by Estimating the Fitness Function. *J.IEEJ C*, 120(1):123–129, 2000.

110. J.J. Grefenstette and C.L. Ramsey. An approach to anytime learning. In *Proc. 9th Int'l Conf. on Machine Learning*, pages 189–195, 1992.

111. M.D. Kidwell and D.J. Cook. Genetic algorithm for dynamic task scheduling. In *Proc. IEEE 13th Int'l Phoenix Conference on Computers and Communications*, pages 61–67, 1994.

112. G. Pipe, T.C. Fogarty, and A. Winfield. Hybrid adaptive heuristic critic architectures for learning in maze with continuous search space. In *Parallel Problem Solving from Nature*, pages 482–491. Springer, 1994.

113. C.L. Ramsey and J.J. Grefenstette. Case-based initialization of genetic algorithms. In *Proc. 5th ICGA*, pages 84–91, 1993.

114. J.J. Grefenstette. Genetic algorithms for changing environments. In *Proc. of Parallel Problem Solving from Nature II*, pages 137–144, 1992.

115. H. G. Cobb. An investigation into the use of hypermutation as an adaptive operator in genetic algorithms having continuous, time-dependent nonstationary environments. Technical report, NRL Memorandum Report 6760, 1990. pp. 523-529.

116. H.G. Cobb and J.J. Grefensttete. Genetic Algorithms for Tracking Changing Environments. In *Proc. 5th ICGA*, pages 84–91, 1993.

117. T. Mori, H. Kita, and Y. Nishikawa. Adaptation to changing environments by Thermo-Dynamical Genetic Algorithms. *Trans.ISCIE*, 12(4):240–249, 1999.

118. D.E. Goldberg and R.E. Smith. Nonstationary function optimization using genetic algorithms with dominance and diploidy. In *Proc. 2nd ICGA*, pages 59–68, 1987.

119. B.S. Hadad and C.F. Eick. Supporting polyploidy in genetic algorithms using dominance vectors. In *6th Int'l. Conf. on Evolutionary Programming*, pages 223–234. Springer, 1997.

120. P. Ng and K.C. Wong. A new diploid scheme and dominance change mechanism for non-stationary function optimization. In *Proc. 6th ICGA*, pages 159–166, 1995.

121. R.E. Smith. Diploid genetic algorithms for search in time varying environments. In *Proc. Annual Southeast Regional Conference of the ACM*, pages 175–179, 1987.

122. C. Ryan. Diploidy without dominance. In *3rd Nordic Workshop on Genetic Algorithms*, pages 63–70, 1997.

123. D. Dasgupta and D. R. McGregor. Nonstationary function optimization using the structured genetic algorithm. In *Proc. of Parallel Problem Solving from Nature II*, pages 145–154, 1992.

124. K. Okura and K. Ueda. Genetic Algorithms for the Optimization of Nonstationary Functions. *Trans.ISCIE*, 8(6):269–276, 1995.

125. K. Mori, M. Tsukiyama, and T. Fukuda. A proposal for an Immune Algorithm with Diversity and its Application to Load Balancing. *Trans. IEEJ*, 113-C(10):872–878, 1993.

126. N. Mori, S. Imanishi, H. Kita, and Y. Nishikawa. Adaptation to a Changing Environment by Means of the memory based thermodynamical Genetic Algorithm. In *Proc. of 7nd ICGA*, pages 299–306, 1998.

127. K. Trojanowski and Z. Michalewicz. Searching for Optima in Non-stationary Environments. In *Proc. of CEC*, pages 1843–1850, 1999.

128. Y. Sano, H. Kita, I. Kaji, and M. Yamaguchi. Application of a genetic algorithm with subpopulations to optimization of dynamic fitness functions. In *SICE 30th Intelligent System Symposium*, pages 97–102, 2003.

129. Y. Sano and H. Kita. Optimization of Noisy Fitness Functions by means of Genetic Algorithms using History of Search with Test of Estimation. In *Proc. CEC 2002*, pages 360–365, 2002.

130. Y. Sano, H. Kita, I. Kamihira, and M. Yamaguchi. Online Optimization of an Engine Controller by means of a Genetic Algorithm using History of Search. In *Proc. SEAL 2000*, pages 2929–2934, 2000.

131. S. Takahashi, H. Kita, H. Suzuki, T. Sudo, and S. Markon. Optimization of the Control of Multi-car Elevators on PC Clusters. In *SICE 15th Autonomous Distributed System Symposium*, pages 1–6, 2003.

132. S. Takahashi. Optimization of Multicar Elevator Control by GA on PC Clusters. Master's thesis, Tokyo Institute of Technology, 2003.

133. Kenneth De Jong, David B. Fogel, and Hans-Paul Schwefel. A history of evolutionary computation. In Th. Bäck, D. B. Fogel, and Z. Michalewicz, editors, *Handbook of Evolutionary Computation*, pages A2.3:1–12. Oxford University Press, New York NY, 1997.

134. Günter Rudolph. A time travel to the early theory of evolution strategies. In T. Bartz-Beielstein, G. Jankord, B. Naujoks, G. Rudolph, and K. Schmitt, editors, *Hans–Paul Schwefel – Festschrift*, pages 85–89. Universität Dortmund, Lehrstuhl für Systemanalyse, 2006.

135. H.-P. Schwefel. Kybernetische evolution als strategie der exprimentellen forschung in der strömungstechnik. Master's thesis, Technical University of Berlin, Germany, 1965.

136. I. Rechenberg. *Evolutionsstrategie: Optimierung technischer Systeme nach Prinzipien der biologischen Evolution.* PhD thesis, Department of Process Engineering, Technical University of Berlin, Germany, 1971.

137. H.-P. Schwefel. *Evolution and Optimum Seeking.* Sixth-Generation Computer Technology. Wiley, New York NY, 1995.

138. Hans-Paul Schwefel, Günter Rudolph, and Thomas Bäck. Contemporary evolution strategies. Interner Bericht der Systems Analysis Research Group SYS–6/95, Fachbereich Informatik, Universität Dortmund, Germany, Dezember 1995.

139. H.-G. Beyer and H.-P. Schwefel. Evolution strategies—A comprehensive introduction. *Natural Computing*, 1:3–52, 2002.

140. H.-P. Schwefel. *Numerische Optimierung von Computer-Modellen mittels der Evolutionsstrategie*, volume 26 of *Interdisciplinary Systems Research*. Birkhäuser, Basel, Switzerland, 1977.

141. Thomas Bartz-Beielstein. Experimental analysis of evolution strategies— overview and comprehensive introduction. Interner Bericht des Sonderforschungsbereichs 531 Computational Intelligence CI–157/03, Universität Dortmund, Germany, November 2003.

142. T. J. Santner, B. J. Williams, and W. I. Notz. *The Design and Analysis of Computer Experiments.* Springer, Berlin, Heidelberg, New York, 2003.

143. Hans-Paul Schwefel. Direct search for optimal parameters within simulation models. In R. D. Conine, E. D. Katz, and J. E. Melde, editors, *Proceedings Twelfth Annual Simulation Symposium, Tampa FL*, pages 91–102, Long Beach CA, 1979. IEEE Computer Society.

144. Farhad Azadivar. Simulation optimization methodologies. In *WSC '99: Proceedings of the 31st Winter Simulation Conference*, pages 93–100, New York NY, 1999. Association for Computing Machinery.

145. J. Banks, J. S. Carson, B. L. Nelson, and D. M. Nicol. *Discrete Event System Simulation.* Prentice Hall, Upper Saddle River NJ, 2001.

146. H.-G. Beyer. Evolutionary algorithms in noisy environments: Theoretical issues and guidelines for practice. *CMAME (Computer Methods in Applied Mechanics and Engineering)*, 186:239–267, 2000.

147. Y. Sano and H. Kita. Optimization of noisy fitness functions by means of genetic algorithms using history of search. In M. Schoenauer et al., editors, *Parallel Problem Solving from Nature (PPSN VI)*, volume 1917 of *Lecture Notes in Computer Science*, pages 571–580, Berlin, Heidelberg, New York, 2000. Springer.

148. Dirk V. Arnold. Evolution strategies in noisy environments—a survey of existing work. In L. Kallel, B. Naudts, and A. Rogers, editors, *Theoretical Aspects of Evolutionary Computing*, pages 239–249. Springer, Berlin, Heidelberg, New York, 2001.

149. J. Branke, C. Schmidt, and H. Schmeck. Efficient fitness estimation in noisy environments. In L. Spector, editor, *Genetic and Evolutionary Computation Conference (GECCO'01)*, pages 243–250, San Francisco CA, 2001. Morgan Kaufmann.

150. Thomas Bartz-Beielstein and Sandor Markon. Tuning search algorithms for real-world applications: A regression tree based approach. In G. W. Greenwood, editor, *Proceedings 2004 Congress on Evolutionary Computation (CEC'04), Portland OR*, volume 1, pages 1111–1118, Piscataway NJ, 2004. IEEE.

151. H.-G. Beyer. *The Theory of Evolution Strategies.* Springer, Berlin, Heidelberg, New York, 2001.

152. R. Jin, X. Du, and W. Chen. The use of metamodeling techniques for optimization under uncertainty. *Journal of Structural & Multidisciplinary Optimization* (in press), 2005.

153. J.E. Chen, C.H. Chen, and D.W. Kelton. Optimal computing budget allocation of indifference-zone-selection procedures. Working paper, taken from `http://www.cba.uc.edu/faculty/keltonwd`. Cited 6 January 2005, 2003.

154. Thomas Bartz-Beielstein, Daniel Blum, and Jürgen Branke. Particle swarm optimization and sequential sampling in noisy environments. In Richard Hartl and Karl Doerner, editors, *Proceedings 6th Metaheuristics International Conference (MIC2005)*, pages 89–94, Vienna, Austria, 2005.

155. Sandor Markon, Dirk V. Arnold, Thomas Bäck, Thomas Beielstein, and Hans-Georg Beyer. Thresholding—A selection operator for noisy ES. In J.-H. Kim, B.-T. Zhang, G. Fogel, and I. Kuscu, editors, *Proceedings 2001 Congress on Evolutionary Computation (CEC'01), Seoul*, pages 465–472, Piscataway NJ, 2001. IEEE.

156. M. Minsky. *The Society of Mind.* Simon and Schuster, New York NY, 1985.

157. Thomas Beielstein and Sandor Markon. Threshold selection, hypothesis tests, and DOE methods. Interner Bericht des Sonderforschungsbereichs 531 Computational Intelligence CI–121/01, Universität Dortmund, Germany, Dezember 2001.

158. Thomas Bartz-Beielstein. *New Experimentalism Applied to Evolutionary Computation.* PhD thesis, Universität Dortmund, Germany, April 2005.

159. J. Kennedy and R.C. Eberhart. Particle swarm optimization. In *Proceedings IEEE International Conference on Neural Networks*, volume IV, pages 1942–1948, Piscataway NJ, 1995. IEEE.

160. J. Kennedy and R.C. Eberhart. *Swarm Intelligence.* Morgan Kaufmann, San Francisco CA, 2001.

161. K.E. Parsopoulos and M.N. Vrahatis. Recent approaches to global optimization problems through particle swarm optimization. *Natural Computing*, 1(2–3):235–306, 2002.

162. K. E. Parsopoulos and M. N. Vrahatis. On the computation of all global minimizers through particle swarm optimization. *IEEE Transactions on Evolutionary Computation*, 8(3):211–224, 2004.

163. R.C. Eberhart and Y. Shi. Comparison between genetic algorithms and particle swarm optimization. In V.W. Porto, N. Saravanan, D. Waagen, and A.E. Eiben, editors, *Evolutionary Programming*, volume VII, pages 611–616. Springer, Berlin, Heidelberg, New York, 1998.

164. Y. Shi and R.C. Eberhart. Empirical study of particle swarm optimization. In P. J. Angeline, Z. Michalewicz, M. Schoenauer, X. Yao, and A. Zalzala, editors, *Proceedings of the Congress of Evolutionary Computation*, volume 3, pages 1945–1950, Piscataway NJ, 1999. IEEE.

165. Thomas Bartz-Beielstein, Marcel de Vegt, Konstantinos E. Parsopoulos, and Michael N. Vrahatis. Designing particle swarm optimization with regression trees. Interner Bericht des Sonderforschungsbereichs 531 Computational Intelligence CI–173/04, Universität Dortmund, Germany, Mai 2004.

166. Thomas Beielstein, Konstantinos E. Parsopoulos, and Michael N. Vrahatis. Tuning PSO parameters through sensitivity analysis. Interner Bericht des Son-

derforschungsbereichs 531 *Computational Intelligence* CI–124/02, Universität Dortmund, Germany, Januar 2002.

167. Thomas Bartz-Beielstein, Konstantinos E. Parsopoulos, and Michael N. Vrahatis. Analysis of particle swarm optimization using computational statistics. In T.-E. Simos and Ch. Tsitouras, editors, *Proceedings International Conference Numerical Analysis and Applied Mathematics (ICNAAM)*, pages 34–37, Weinheim, Germany, 2004. Wiley-VCH.

168. M. Clerc and J. Kennedy. The particle swarm-explosion, stability, and convergence in a multidimensional complex space. *IEEE Transactions on Evolutionary Computation*, 6(1):58–73, 2002.

169. J. Kennedy. Bare bones particle swarms. In *Proceedings 2003 IEEE Swarm Intelligence Symposium*, pages 80–87, Piscataway NJ, 2003. IEEE.

170. Y. Shi. Particle swarm optimization. *IEEE CoNNectionS – The Newsletter of the IEEE Neural Networks Society*, 2(1):8–13, February 2004.

171. K. E. Parsopoulos and M. N. Vrahatis. Particle swarm optimizer in noisy and continuously changing environments. In M.H. Hamza, editor, *Artificial Intelligence and Soft Computing*, pages 289–294. IASTED/ACTA Press, 2001.

172. Thiemo Krink, Bogdan Filipic, Gary B. Fogel, and Rene Thomsen. Noisy optimization problems - a particular challenge for differential evolution? In *Proceedings of the 2004 IEEE Congress on Evolutionary Computation*, pages 332–339, Portland OR, 20-23 June 2004. IEEE.

173. Ingo Rechenberg. *Evolutionsstrategie. Optimierung technischer Systeme nach Prinzipien der biologischen Evolution.* frommann–holzboog, Stuttgart, Germany, 1973.

174. F. Glover. Tabu Search – Part I. *ORSA Journal on Computing, Vol.1, No.3*, pages 190–206, 1989.

175. H. Gould. *Thermal and Statistical Physics Simulations.* John Wiley, New York NY, 1995.

176. B. Chopard and M. Droz. *Cellular Automata Modeling of Physical Systems.* Cambridge U.P., 1998.

177. R. A. Howard. *Dynamic Programming and Markov Processes.* MIT Press, 1960.

178. L. P. Kaelbling. *Learning in Embedded Systems.* MIT Press, 1993.

179. H.J. Kushner and D.S. Clark. *Stochastic Approximation Methods for Constrained and Unconstrained Systems*, volume 26 of *Applied Mathematical Sciences.* Springer-Verlag, 1978.

180. T. Beielstein and S. Markon. Threshold Selection, Hypothesis Tests and DOE Methods. In *Proc. Congress on Evolutionary Computation (CEC'2002, Honolulu, HI)*, volume 1, pages 777–782. IEEE, 2002.

181. S. Markon, H.Kita, and Y. Nishikawa. Adaptive Optimal Elevator Group Control by Neural Networks. *Proc. 2nd Symp. Jap. Neural Network Soc.*, pages 187–188, 1991.

182. S. Markon, H.Kita, and Y. Nishikawa. Reinforcement Learning for Stochastic System Control by using a Feature Extraction with BP Neural Networks. *IEICE NC91-126*, pages 209–214, 1991.

183. D.E. Rumelhart, J.L. McClelland, and the PDP Research Group. *Parallel Distributed Processing.* The MIT Press, 1986.

184. A. Takaoka. *Development of a Neural Network Simulator with Linked Weights and Conjugate Gradient Learning Rule, M.Sc. Thesis.* Kyoto University, Kyoto, 1990.

185. J. Kiefer and J. Wolfowitz. Stochastic Estimation of the Maximum of a Regression Function. *Ann. Math. Stat.* *23*, pages 462–466, 1975.
186. S. Yakowitz and E. Lugosi. Random Search in the Presence of Noise with Application to Machine Learning. *SIAM J. Sci. Stat. Comput. 11, No.4*, pages 702–712, 1990.
187. R.Y. Rubinstein and A. Shapiro. *Discrete Event Systems*. John Wiley, 1993.
188. H.-P. Schwefel, I. Wegener, and K. Weinert, editors. *Advances in Computational Intelligence—Theory and Practice*. Springer, Berlin, Heidelberg, New York, 2003.
189. A. E. Eiben and J. E. Smith. *Introduction to Evolutionary Computing*. Springer, Berlin, Heidelberg, New York, 2003.
190. Thomas Bartz-Beielstein. *Experimental Research in Evolutionary Computation—The New Experimentalism*. Springer, Berlin, Heidelberg, New York, 2006.
191. Jörn Mehnen, Thomas Michelitsch, Thomas Bartz-Beielstein, and Christian W. G. Lasarczyk. Multiobjective evolutionary design of mold temperature control using DACE for parameter optimization. In H. Pfützner and E. Leiss, editors, *Proceedings Twelfth International Symposium Interdisciplinary Electromagnetics, Mechanics, and Biomedical Problems (ISEM 2005)*, volume L11-1, pages 464–465, Vienna, Austria, 2005. Vienna Magnetics Group Reports.
192. Klaus Weinert, Jörn Mehnen, Thomas Michelitsch, Karlheinz Schmitt, and Thomas Bartz-Beielstein. A multiobjective approach to optimize temperature control systems of moulding tools. *Production Engineering Research and Development, Annals of the German Academic Society for Production Engineering*, XI(1):77–80, 2004.
193. Jörn Mehnen, Thomas Michelitsch, Thomas Bartz-Beielstein, and Nadine Henkenjohann. Systematic analyses of multi-objective evolutionary algorithms applied to real-world problems using statistical design of experiments. In R. Teti, editor, *Proceedings Fourth International Seminar Intelligent Computation in Manufacturing Engineering (CIRP ICME'04)*, volume 4, pages 171–178, Naples, Italy, 2004.
194. Thomas Bartz-Beielstein and Boris Naujoks. Tuning multicriteria evolutionary algorithms for airfoil design optimization. Interner Bericht des Sonderforschungsbereichs 531 Computational Intelligence CI–159/04, Universität Dortmund, Germany, Februar 2004.
195. Thomas Bartz-Beielstein, Mike Preuß, and Sandor Markon. Validation and optimization of an elevator simulation model with modern search heuristics. In T. Ibaraki, K. Nonobe, and M. Yagiura, editors, *Metaheuristics: Progress as Real Problem Solvers*, Operations Research/Computer Science Interfaces, pages 109–128. Springer, Berlin, Heidelberg, New York, 2005.
196. Thomas Bartz-Beielstein, Christian Lasarczyk, and Mike Preuß. Sequential parameter optimization. In B. McKay et al., editors, *Proceedings 2005 Congress on Evolutionary Computation (CEC'05), Edinburgh, Scotland*, volume 1, pages 773–780, Piscataway NJ, 2005. IEEE Press.
197. Marcel de Vegt. Einfluss verschiedener Parametrisierungen auf die Dynamik des Partikel-Schwarm-Verfahrens: Eine empirische Analyse. Interner Bericht der Systems Analysis Research Group SYS–3/05, Universität Dortmund, Fachbereich Informatik, Germany, Dezember 2005.
198. Marko Tosic. Evolutionäre Kreuzungsminimierung. Diploma thesis, University of Dortmund, Germany, January 2006.

199. Thomas Bartz-Beielstein. Evolution strategies and threshold selection. In M. J. Blesa Aguilera, C. Blum, A. Roli, and M. Sampels, editors, *Proceedings Second International Workshop Hybrid Metaheuristics (HM'05)*, volume 3636 of *Lecture Notes in Computer Science*, pages 104–115, Berlin, Heidelberg, New York, 2005. Springer.

200. Thomas Bartz-Beielstein, Konstantinos E. Parsopoulos, and Michael N. Vrahatis. Design and analysis of optimization algorithms using computational statistics. *Applied Numerical Analysis & Computational Mathematics (ANACM)*, 1(2):413–433, 2004.

201. F. Pukelsheim. *Optimal Design of Experiments*. Wiley, New York NY, 1993.

202. N. R. Draper and H. Smith. *Applied Regression Analysis*. Wiley, New York NY, 3rd edition, 1998.

203. D. C. Montgomery. *Design and Analysis of Experiments*. Wiley, New York NY, 5th edition, 2001.

204. J. Sacks, W. J. Welch, T. J. Mitchell, and H. P. Wynn. Design and analysis of computer experiments. *Statistical Science*, 4(4):409–435, 1989.

205. D.R. Jones, M. Schonlau, and W.J. Welch. Efficient global optimization of expensive black-box functions. *Journal of Global Optimization*, 13:455–492, 1998.

206. S.N. Lophaven, H.B. Nielsen, and J. Søndergaard. DACE—A Matlab Kriging Toolbox. Technical Report IMM-REP-2002-12, Informatics and Mathematical Modelling, Technical University of Denmark, Copenhagen, Denmark, 2002.

207. E. H. Isaaks and R. M. Srivastava. *An Introduction to Applied Geostatistics*. Oxford University Press, Oxford, U.K., 1989.

208. S.N. Lophaven, H.B. Nielsen, and J. Søndergaard. Aspects of the Matlab Toolbox DACE. Technical Report IMM-REP-2002-13, Informatics and Mathematical Modelling, Technical University of Denmark, Copenhagen, Denmark, 2002.

209. W. J. Welch, R. J. Buck, J. Sacks, H. P. Wynn, T. J. Mitchell, and M. D. Morris. Screening, predicting, and computer experiments. *Technometrics*, 34:15–25, 1992.

210. M. Schonlau. *Computer Experiments and Global Optimization*. PhD thesis, University of Waterloo, Ontario, Canada, 1997.

211. M. D. McKay, R. J. Beckman, and W. J. Conover. A comparison of three methods for selecting values of input variables in the analysis of output from a computer code. *Technometrics*, 21(2):239–245, 1979.

212. R. Aslett, R. J. Buck, S. G. Duvall, J. Sacks, and W. J. Welch. Circuit optimization via sequential computer experiments: design of an output buffer. *Journal of the Royal Statistical Society: Series C (Applied Statistics)*, 47(1):31–48, 1998.

213. Hans-Georg Beyer. An alternative explanation for the manner in which genetic algorithms operate. *BioSystems*, 41:1–15, 1997.

214. D. G. Mayo. *Error and the Growth of Experimental Knowledge*. The University of Chicago Press, Chicago IL, 1996.

215. Hyper-Building Research Group. *Report of the Hyper-Building Research*. Hyper-Building Research Group, 1996. in Japanese.

216. K. Mimaki, S. Markon, H. Kita, Y. Komoriya, and Y. Nishikawa. Modeling and Analysis of Complex Traffic in Buildings. *Proc. IEEE SMC'99*, 1999.

217. R.G. Smith. The Contract Net Protocol: High-level Communication and Control in a Distributed Problem Solve. *IEEE Trans. Comput*, C-29(12):1104–1113, 1980.

218. H. Sato, I. Ono, and S. Kobayashi. A Proposal and Evaluation of a Generation Change Model for Genetic Algorithms. *J.AIJ*, 12:734–744, 1998.

219. T. Sudo and S. Markon. The Performance of Multi-Car Linear Motor Elevator. *Elevator Technology 11, Proc. Of ELEVCON 2001*, pages 141–149, 2001.

220. T. Sudo, H. Suzuki, S. Markon, and H. Kita. Effectiveness and control strategies of multi-car elevators for high-rise buildings. *TRANSLOG'02*, 2002.

221. N.G. Hall. *Operational Research Techniques for Robotic Systems*. John Wiley, N. Y., 1998.

222. D.S. Kim, S.Y. Jang, and W.Y. Lee. A Study on the FMS Scheduling Method Considering AGV Request Time Using Simulation Technique. *INFORMATION KORMS, Seoul2000 Korea*, pages 1097–1103, 2000.

223. H. Kise, G. Hu, and C. Xie. Tabu-based Optimization for Input/Output Scheduling in Automated Warehouses. *Proceedings (CD-ROM), the sixth Metaheuristics International Congress, Vienna, Austria, August 22-25, 2005*, pages 33–38, 2005.

224. J.P. van den Berg. A literature survey on planning and control of warehousing systems. *IEE Transactions*, 31:751–762, 1999.

225. R. de Koster and E. van der Poor. Routing orderpickers in a warehouse: a comparison between optimal and heuristic solutions. *IEE Transactions*, 30:469–480, 1998.

226. D.H. Ratliff and A.S. Rosenthal. Order picking in a rectangular warehouse: A solvable case of the traveling salesman problem. *Operations Research*, 31:507–521, 1983.

227. F. Glover and M. Laguna. Tabu search. In *Modern heuristic techniques for combinatorial problems*, pages 70–143. Wiley, 1993.

Index

Other titles published in this Series (continued):

Analysis and Control Techniques for Distribution Shaping in Stochastic Processes
Michael G. Forbes, J. Fraser Forbes, Martin Guay and Thomas J. Harris
Publication due August 2006

Process Control Performance Assessment
Andrzej Ordys, Damien Uduehi and Michael A. Johnson (Eds.)
Publication due August 2006

Adaptive Voltage Control in Power Systems
Giuseppe Fusco and Mario Russo
Publication due September 2006

Distributed Embedded Control Systems
Matjaž Colnarič, Domen Verber and Wolfgang A. Halang
Publication due October 2006

Modelling and Analysis of Hybrid Supervisory Systems
Emilia Villani, Paulo E. Miyagi and Robert Valette
Publication due November 2006

Model-based Process Supervision
Belkacem Ould Bouamama and Arun K. Samantaray
Publication due February 2007

Continuous-time Model Identification from Sampled Data
Hugues Garnier and Liuping Wang (Eds.)
Publication due May 2007

Magnetic Control of Tokamak Plasmas
Marco Ariola and Alfredo Pironti
Publication due May 2007

Process Control
Jie Bao, and Peter L. Lee
Publication due June 2007

Optimal Control of Wind Energy Systems
Iulian Munteanu, Antoneta Iuliana Bratcu, Nicolas-Antonio Cutululis and Emil Ceanga
Publication due November 2007